Dynamic Response and Deformation Characteristic of Saturated Soft Clay under Subway Vehicle Loading

Springer Environmental Science and Engineering

For further volumes:
http://www.springer.com/series/10177

Yiqun Tang · Jie Zhou · Xingwei Ren · Qi Yang

Dynamic Response and Deformation Characteristic of Saturated Soft Clay under Subway Vehicle Loading

Yiqun Tang
Department of Geotechnical Engineering
Tongji University, Shanghai
People's Republic of China

Jie Zhou
Department of Geotechnical Engineering
Tongji University, Shanghai
People's Republic of China

Xingwei Ren
Department of Geotechnical Engineering
Tongji University, Shanghai
People's Republic of China

Qi Yang
Department of Geotechnical Engineering
Tongji University, Shanghai
People's Republic of China

Funded by National Natural Science Foundation of China (Grant No. 40372124 and No. 41072204)

ISSN 2194-3214   ISSN 2194-3222 (electronic)
ISBN 978-3-642-41986-7   ISBN 978-3-642-41987-4 (eBook)
DOI 10.1007/978-3-642-41987-4
Springer Heidelberg New York Dordrecht London

Library of Congress Control Number: 2013957571

© Science Press Ltd, Beijing and Springer-Verlag Berlin Heidelberg 2014
This work is subject to copyright. All rights are reserved by the Publishers, whether the whole or part of the material is concerned, specifically the rights of translation, reprinting, reuse of illustrations, recitation, broadcasting, reproduction on microfilms or in any other physical way, and transmission or information storage and retrieval, electronic adaptation, computer software, or by similar or dissimilar methodology now known or hereafter developed. Exempted from this legal reservation are brief excerpts in connection with reviews or scholarly analysis or material supplied specifically for the purpose of being entered and executed on a computer system, for exclusive use by the purchaser of the work. Duplication of this publication or parts thereof is permitted only under the provisions of the Copyright Law of the Publishers' locations, in its current version, and permission for use must always be obtained from Springer. Permissions for use may be obtained through RightsLink at the Copyright Clearance Center. Violations are liable to prosecution under the respective Copyright Law.
The use of general descriptive names, registered names, trademarks, service marks, etc. in this publication does not imply, even in the absence of a specific statement, that such names are exempt from the relevant protective laws and regulations and therefore free for general use.
While the advice and information in this book are believed to be true and accurate at the date of publication, neither the authors nor the editors nor the publishers can accept any legal responsibility for any errors or omissions that may be made. The publishers make no warranty, express or implied, with respect to the material contained herein.

Printed on acid-free paper

Springer is part of Springer Science+Business Media (www.springer.com)

Foreword

Rapid development of the Chinese economy has spurred and accelerated major urbanization construction. With lots of people migrating into large and medium cities, the previous existing urban transportation hardly meets the demand of the population expansion and urban sprawl. Establishing spacing transportation network is the key to alleviate the situation of traffic congestion, in which subway plays a significant role to develop the traffic underground space without any large disturbance to the surface construction. Subway systems bring free-flow traffic at all times and realize multidimensional space traffic to make faster modernization of urban development.

In China, since the first subway showed up in Beijing in the 1960s, Shanghai, Guangzhou, Shenzhen, Nanjing and Tianjin all followed and interlaced the heartland by subways. Currently more than 20 cities have also planned or already opened subway construction to traffic such as Wuhan, Zhengzhou, Changsha, Suzhou, and Ningbo. Particularly, in Beijing, Shanghai, and Guangzhou, in these three big cities, subway systems all develop so well that about 200–400 km subway line has been established to form a really huge traffic hub, which promote the urban development greatly.

Subway construction is really helpful for solving urban traffic problems, but a number of issues also should be paid much attention especially in coastal cities with large area of soft soil located. In poor geological conditions, the study for subway design and construction is essential to prevent and reduce the occurrence of accidents. Meantime, there is no specific and detailed study before the subway system started construction. It is much lack of long-time experiments and monitoring during the fast construction of underground traffic space. Hence, without thorough understanding of geological conditions and practice experiences, several hidden problems and issues would be resulted in many aspects.

This monograph has been prepared with the combined effort of all researchers in the group under Prof. Tang's leading. It is the greatest achievement on several years' research accumulation. Lots of results about subway construction were figured out by authors especially in soft soil area. It is very valuable research results from the practice experiences of Shanghai Subway Lines 1 and 2 and also some kind of exploratory study for the subway construction in Shanghai or other soft soil areas.

In recognition of the importance of improving and expanding the theoretical base to provide reference for subway design and construction, Prof. Tang has been conducting comprehensive research supported by the National Natural Science Foundation of China (Grant No. 40372124 and No. 41072204) and Shanghai Post-doctoral Foundation. The dynamic response characteristics of soft clay surrounding tunnel under subway vibration load are studied in this research project. The main content is characterized as follows:

1. From a specific view, the soil structure was selected as a start point of research in depth. Through field monitoring and laboratory experiments, large numbers of data were provided as analysis basis to the potential effects. Considering advantages and disadvantages in subway construction, interior characteristics of soil structure, bearing capacity, and compressibility and settlement problems were all analyzed.
2. The soil deformation and dynamic response effect were discussed along with the intrinsic properties of soil with the vibration load. It includes the soil response frequency to subway vibration load, soil dynamic characteristics under high-frequency and low frequency load.
3. The concept of multi-methodology of geological monitoring and various testing and vibration experiments is presented in this book. Numerous data were summarized and were mutually proven by each other to establish relevant mathematical model. It can be suitable for field analyzing and also has significant application.
4. The engineering effect on surrounding environment and buildings was discussed based on the long-term subway foundation, soil characteristics, and the vibration load state. Some valuable prediction information can be provided.

The research results in this monograph have great practice value to guide the subway design and construction projects, particularly for effectively controlling and alleviating or even preventing the land subsidence effects caused by large area subway vibration load. Being so systematic and practical, this monograph is unique and outstanding on subway construction since the subway system developed in large scale in the 1990s in Shanghai. It's really hard to find another similar one even all over China. I am pretty sure that once this monograph is published, it must be a great reference and have a significant guide on underground space network traffic construction in Shanghai, Eastern China, and all the southern soft soil areas.

Chinese Academy of Engineering
Beijing, People's Republic of China
May 30th 2013

Academician Yaoru Lu

Preface

Rapid development of the Chinese economy has spurred and accelerated major urbanization construction. Traffic problems emerge gradually. Establishing spacing transportation network is the key to alleviate the situations of large population, dense buildings, narrow streets, and traffic congestion. Subway system is paid more attentions due to its advantages of high speed, safety and comfortability. Most importantly, it develops the traffic underground space without any large disturbance to the surface construction. Currently, subway lines are constructed in Shanghai, Beijing, Tianjin, Shenzhen, Ningbo, and Chongqing to mitigate the traffic loads. Even some small cities are applying for subway construction now. As for Shanghai, there are already 400 km lines in total in the rail transit system. Further planning to 2020, there will be 21 lines in Shanghai Metro, which is 840 km in total. However, at the same time, this sharp development of subway systems has brought in lots of environmental problems as well. Particularly, large deformation arises in the surrounding soil near subway tunnel and the foundation. Ground subsidence influences the surrounding environment.

Supported by the National Natural Science Foundation of China (Grant No. 40372124 and No. 41072204), National Specialty Comprehensive Reform Pilot Project (the project number: higher educational letter [2013] No. 2), National Key Technologies R&D Program of China (Grant No. 2012BAJ11B04), and Shanghai Postdoctoral Foundation, this monograph comprehensively presents the research on the dynamic response characteristics of saturated soft clay surrounding the subway tunnel under the subway vibration loads. Most subway lines in Shanghai were or are being designated in the saturated soft clay layer, which is characterized by high void ratio, large natural water content, poor permeability, high compressibility and low shear strength, etc. Under long-term subway vibration loading, the saturated soft clay may gradually be compressed and residual deformation will result in deflection of tunnel axis, land subsidence, and even tilt or collapse of surrounding buildings. To prevent such disasters, the dynamic properties, deformation mechanism, and settlement characteristics should be figured out under vibration loading. Aimed at this, during the research process, field sampling, monitoring, and laboratory experiments are combined; elastoplastic theory, material mechanics, fracture mechanics, and nonlinear dynamic mechanics are employed as the basis; dynamic triaxial system (GDS), scanning electron microscopy (SEM), and mercury intrusion porosimetry (MIP) apparatus are utilized to simulate the soil dynamic properties. Through all the above, the microstructure deformation and failure mechanism are analyzed; meanwhile, the long-term settlement and land subsidence characteristics are simulated by finite element method (FEM); an applicable prediction model is proposed. Some valuable conclusions are drawn finally.

Chapter 1 is Introduction. It summarizes the recent researches and progresses all over the world of the study on five areas including soil structure, the dynamic response to subway vibration loads, the dynamic properties of soft clay under vibration loads, microstructure of soft clay, and the long-term settlement of soft foundation under vibration loads. Hereby it points out the purpose and research strategy in this monograph.

Chapter 2 is Field Tests. All the tests were conducted around the site of Jing'an Temple Station in Line 2, Shanghai Metro. There were 5 boreholes drilled along and perpendicular to the tunnel axis. Earth pressure and pore water pressure transducers were placed in different distances away from the tunnel edge along and vertically at various depths. The developments of earth pressure and pore water pressure under subway vibration loads were monitored in real time. Processing the field data, it was found that the frequency of soft clay responding to train operation can be divided into two parts: high frequency (2.4–2.6 Hz) and low frequency (0.4–0.6 Hz). The attenuation relationship of the dynamic response along the distance in the vertical line of the tunnel axis is also derived. Through this formula, the impact scope and the dynamic response values can be calculated, and the influence of vibrated subway loading on the surrounding buildings can be evaluated and predicted. All these can provide theoretical reference for the design and construction of the buildings around subway.

Chapter 3 is the Laboratory Tests. With the data of field monitoring, by means of laboratory tests of CKC, GDS (Global Digital Systems), SEM (scanning electron microscopy), and MIP (mercury intrusion porosimetry), the further study on the variation of pore water pressure, soil strength, dynamic constitutive relation, and dynamic elastic modulus under different vibration frequencies and vibration cycle numbers are presented. In addition, the microstructure deformation and damage mechanism are also discussed. It is resulted that the increase of pore water pressure is divided into three stages. They are rapid growth stage, slow growth stage, and stable stage. Logistic model is fitted for the variation of pore water pressure, in which the correlation coefficient reaches above 0.99. During the rapid growth stage, the velocity of increase on pore water pressure is not a stable value. The pore water pressure curve is a sharp dip line during the initial short time and then gets into slow decaying. Through regression analysis, the variation of pore water pressure is much consistent with the ExpDecay2 model line. In addition, after the cessation of the vibrated subway loading applied, the pore water pressure sharply declines and then gets into a stable value a little above the hydrostatic pressure. Secondly, the deformation of saturated soft clay soil under the cyclic subway loading is much related to the cyclic stress ratio, confining pressure, vibration frequency, and number of vibration. There exists a threshold cyclic stress ratio value during the vibrated loading, which is associated with the properties of soil, the vibrated loading, the confined value, etc. When the vibrated loading applied is smaller than the value under the threshold cyclic stress ratio, with the increasing of the cyclic number, the deformation of soil gets larger but the rate declines and the amplitude of vibration gradually reduces till to a constant. Vice versa, as the cyclic number is growing, the deformation breaks out quickly and is ultimately destroyed. While the axial deformation in the bottom of subway tunnel lasts a very short rebound phase then immediately gets into the plastic deformation stage and large axial deformation occurs. Hence, even though large deformation will not take place in the surrounding soil at a long time in the subway operation, the differential settlement may generate as time goes on.

Chapter 4 is the Research of Microstructure. By means of the advanced microtesting methods of SEM (scanning electron microscopy) and MIP (mercury intrusion porosimetry), the variation of microstructure of the saturated soft clay soil under the vibrated subway loading is discussed from qualitative and quantitative point combined. Meanwhile, the deformation mechanism is analyzed in microstructure aspect through the test results. With the automatic mercury intrusion porosimetry, the samples before and after vibration are quantitatively analyzed in pore size distribution, quantity, and other parameters of pore structure. Supplemented by SEM images in the qualitative analysis, the contact of microstructure and the corresponding macroscopic mechanical properties are established. The results show that there is mainly flocculation, cellular-flocculation structure, in the saturated soft clay. The clay minerals are mostly illite with some chlorite, montmorillonite, etc. Structure unit is generally in the shape of thin sheet. The aggregates are flocculent and feathery. The contact status of the structure units appears in the mode of side-surface or side-side, which is resulted in the aerial structure of the muddy clay with high void ratio. The pore size is distributed mainly in the large pore, and the ink-bottle effect shows up during the process of mercury intrusion. When the depth deepens, the cyclic stress ratio increases, and the total specific surface area of pores firstly declines and then rebounds slightly; the uniform pore size, mercury-retention coefficient, and porosity all increase and then slightly reduce. In addition, it is detected that the Masonry Model is not suitable for the explanation of the deformation mechanism of the saturated soft clay under low stress condition. The fractal theory is confirmed to analyze the pore size distribution. When using different fractal models, the description on the pore size distribution sometimes has a little difference. In this book, the fractal results in thermodynamic relation is much better than that in Menger model as for saturated soft clay.

Chapter 5 is the Finite Element Modeling. According to the field monitoring data, three-dimensional finite element computer model is established to simulate the dynamic response of the surrounding soil when the subway passes by. The further long-term land subsidence is also obtained in the subway operation.

Chapter 6 is the Settlement Prediction of Soils Surrounding Subway Tunnel. By means of Newton's quadratic interpolation polynomial method, non-isochronous data sequence is converted to isochronous data sequence. The GM (N, 1) model of non-isochronous data sequence is established. This advantage of using this method is that the impact factors are not simply superposed. Based on the field monitoring data, comparatively analyzing the settlement data, the results show that GM (2, 1) and GM (1, 1) have the higher accuracy. It is much more suitable for the sample volume larger than 4, and the prediction value is much closer to the real settlement. Analyzing the field monitoring data, the rate of settlement in the first half year is obviously lower than that in the second half year.

Chapter 7 is Conclusions and Prospects. This part comprehensively summarizes the research conclusions. Several controversial issues are discussed and then the further research work and prospects are simply described.

This monograph has been prepared with the combined effort of all researchers in the group under Prof. Yiqun Tang's leading, in which Prof. Nianqing Zhou, associate Prof. Ping Yang, Doctor Xi Zhang, Ph.D. student Jie Zhou, Xingwei Ren, and some other master students all have involved in this comprehensive research work in this monograph. Especially, the writing and finalized editing are conducted by Prof. Tang, graduate student Jie Zhou, and Qi Yang. Xingwei Ren, Jun Li, and Qi Yang are also involved in the information processing and interpretation work.

Though this monograph has been published out on our researches, it is just the first step in our research work. A lot of relevant problems are still needed to explore. All the authors hope this monograph can bring in many more researchers' interests to get involved in this area.

Tongji University   Prof. Yiqun Tang
Shanghai, People's Republic of China
June 2013

Contents

1 Introduction 1
   1.1 Research Status 3
      1.1.1 The Study on Soil Structure 3
      1.1.2 The Study on Vehicle Vibration Loading 3
      1.1.3 Dynamic Response of Vehicle Vibration Load 6
      1.1.4 Dynamic Properties of Soft Clay Soil Under
            Dynamic Loading 10
      1.1.5 Microstructure of Soft Clay Soil 13
      1.1.6 The Long-Term Settlement of Soft Clay
            Foundation Under Vehicle Vibration Loading 16
   1.2 Research Content and Methodology 18
      1.2.1 Dynamic Response of Soft Clay Surrounding
            Subway Tunnel to Subway Traffic Loads 19
      1.2.2 Dynamic Characteristics of Soft Clay Under
            Subway Vibration Loads 19
      1.2.3 Microstructure Study of Soft Clay Under Subway
            Vibration Loads 19
      1.2.4 Numerical Simulation 20
      1.2.5 Settlement Prediction of Soil Surrounding
            Subway Tunnel Under Subway Vibration Loads 20
   References 20

2 Field Tests 25
   2.1 Introduction 25
   2.2 Engineering Geology of Soft Soil in Shanghai 26
   2.3 Program Design 28
      2.3.1 Selection of Testing Site 28
      2.3.2 Selection of Testing Instruments 28
      2.3.3 Stratigraphic Section and Instrument Installation 29

	2.4	Results and Analysis	29
		2.4.1 Soil Response Frequency	30
		2.4.2 The Attenuation of Dynamic Response in Perpendicular Direction to Subway Tunnel Axis	32
		2.4.3 The Dynamic Response Development Along the Depth	34
		2.4.4 Development of Pore Pressure	35
		2.4.5 Mechanism Analysis of Pore Pressure Development Under Subway Traffic Loading	36
	2.5	Chapter Summary	38
		References	38
3	**Laboratory Tests**	41
	3.1	Introduction	41
	3.2	Pore Pressure Development	42
		3.2.1 Experimental Summary	43
		3.2.2 The Pore Water in Soft Clay	49
		3.2.3 The Pore Water Development Properties	53
		3.2.4 The Pore Water Pressure Attenuation Properties	58
		3.2.5 The Mechanism Analysis of Influence of Subway Vibration Loads on Pore Water Pressure	59
	3.3	Deformation Characteristics of Saturated Soft Soils	60
		3.3.1 Experimental Introduction	61
		3.3.2 The Factors on Deformation Under Cyclic Loads	65
		3.3.3 Summary	70
	3.4	Strength Characteristics of Saturated Soft Soils	71
	3.5	Creep Behavior of Soil Under Cyclic Loading (Tang et al. 2010)	72
		3.5.1 Test Apparatus and Samples	74
		3.5.2 Test Control Parameter and Procedures	74
		3.5.3 Test Procedures	75
		3.5.4 Composition of Clayey Creep Under Cyclic Loading	76
		3.5.5 The Change of Reversible Elastic Strain of Saturated Clay	81
		3.5.6 Variation of Accumulated Plastic Strain of Saturated Clay	83
		3.5.7 Variation of the Residual Pore Water Pressure	85
		3.5.8 Summaries	86
	3.6	Dynamic Stress-Strain Relationships	87
	3.7	Effective Principal Stress Variation	90
		3.7.1 Experimental Introduction	90
		3.7.2 The Variation of Effective Primary Stress with Loading Time	92
		3.7.3 The Variation of Shear Wave Velocity	95
	3.8	Mechanism Analysis	95
	3.9	Chapter Summary	97
		References	98

4	**Research of Microstructure**		101
	4.1 Introduction		101
	4.2 Qualitative Analysis		104
		4.2.1 Research Method	106
		4.2.2 Preparation of Samples	107
		4.2.3 Basic Characteristics of the Soil Samples	108
		4.2.4 Qualitative Analysis of Observed Result	109
	4.3 Quantitative Analysis		126
		4.3.1 Research Method	127
		4.3.2 Fractal Parameters of MIP Results	131
		4.3.3 The Characteristics of Pore Variation in Undisturbed Soft Clay During the Process of Mercury Intrusion	132
		4.3.4 The Characteristics of Pore Variation in Undisturbed Soft Clay During the Process of Mercury Extrusion	135
		4.3.5 The Variation of Pore Structure Characteristic Parameters in Saturated Soft Clay Under Vibration Loading	136
		4.3.6 Fractal Model and Fractal Dimension	140
	4.4 Correlation Analysis Between Microstructure and Macroscopic Deformation Properties of Saturated Soft Clay Under Subway Loading		147
		4.4.1 Correlation Analysis of Pore Microstructure Parameter and Macroscopic Force	147
		4.4.2 Correlation Analysis of Pore Microstructure Parameters and Macroscopic Deformation	149
	4.5 Chapter Summary		151
	References		152
5	**Finite Element Modeling**		155
	5.1 Introduction		155
	5.2 Finite Element Analysis Software		156
		5.2.1 Numerical Method	156
		5.2.2 The Finite Element Method and ANSYS	156
	5.3 Theoretical Analysis		157
		5.3.1 General Equations of the Dynamic Analysis	158
		5.3.2 Dynamic Finite Element Analysis	161
		5.3.3 Static-Dynamic Analysis of Subway-Soil System	166
	5.4 Simulation of the Dynamic Loading Induced by the Subway Train		169
		5.4.1 Generation Mechanism	170
		5.4.2 Simplified Model for Subway Train	173
		5.4.3 The Simulation of Subway Train Loading	173

	5.5	Model Development	177
		5.5.1 Introduction to the Model	177
		5.5.2 Mesh Generation in FEM Model............................	179
		5.5.3 Material Parameters	184
	5.6	Viscoelastic Artificial Boundary	184
	5.7	The Time Integral Step	186
	5.8	The Results of Transient Analysis....................................	186
	5.9	Chapter Summary	189
		References	190
6	**Settlement Prediction of Soils Surrounding Subway Tunnel**	191
	6.1	Introduction	191
	6.2	Analysis on Subway Tunnel Settlement During Tunneling in Soft Soil..	192
		6.2.1 Factors	192
		6.2.2 Scope of Settlement and Settlement Tank	194
		6.2.3 Relationship Between Settlement and Grouting Quantity	195
		6.2.4 Estimation of the Ground Pre-settlement	198
		6.2.5 Summary of Early Settlement's Analysis	200
	6.3	Analysis of Long-Term Settlement of Subway Tunnel in Soft Soil ..	200
		6.3.1 Factors Affecting Long-Term Settlement of Subway Tunnel in Soft Soil	200
		6.3.2 Mechanism Analysis of Long-Term Additional Settlement in Foundation Caused by Subway Load...........	202
	6.4	Applications of Gray Prediction Theory in Predicting Long-Term Settlement of Subway Tunnel	203
		6.4.1 Research Backgrounds and Introduction	203
		6.4.2 Methodology of Gray Prediction Theory	206
		6.4.3 Model Checking	209
		6.4.4 Applications of Prediction Model............................	210
		6.4.5 Discussions and Error Analyses...............................	213
	6.5	Effective Stress Analysis of Long-Term Settlement in Tunnel	219
	6.6	Chapter Summary	221
		References	222
7	**Conclusions and Prospects**...	225
	7.1	Conclusions	225
	7.2	Prospects for Further Study..	228

Appendix: Major Published Works of the Book Author 229

Index 239

List of Figures

Fig. 2.1	The typical geological strata of Shanghai	27
Fig. 2.2	Plan view of the testing site	28
Fig. 2.3	Schematic diagram of dynamic monitoring system	28
Fig. 2.4	Stratigraphic profile and instrument installation	29
Fig. 2.5	The subway train (Model AC02) diagram of Shanghai Metro Line 2	30
Fig. 2.6	The soil vibration response wave	31
Fig. 2.7	Frequency histogram in monitoring boreholes	32
Fig. 2.8	Attenuation of lateral stress perpendicular to tunnel axis	33
Fig. 2.9	Soil dynamic response amplitude with the depth	34
Fig. 2.10	Response curve of pore water pressure at subsurface 11.5 m	35
Fig. 2.11	Curves of over-static pore water pressure and vibration number at a depth of 11.5 m	36
Fig. 2.12	Curves of over-static pore water pressure and vibration number at a depth of 8.5 m	36
Fig. 2.13	Curves of over-static pore water pressure and vibration number at a depth of 13.5 m	37
Fig. 3.1	GDS dynamic triaxial test system	43
Fig. 3.2	Operation schematic diagram of GDS	44
Fig. 3.3	Example of data analysis module of GDS	45
Fig. 3.4	Preparation workflow before starting test. (**a**) Specimen cutting. (**b**) Filter attaching. (**c**) Latex filming. (**d**) Specimen fixing. (**e**) Water injection for pressure chamber. (**f**) Starting the test	47
Fig. 3.5	GUI of GDS system	48
Fig. 3.6	Electric double layer	50
Fig. 3.7	The variation of potential in relation to distance	51
Fig. 3.8	Schematic diagram of bound water on the clay particle surface	52
Fig. 3.9	Relationship between pore water pressure and time: (**a**) 8.5 m; (**b**) 11.5 m; (**c**) 13.5 m	54

Fig. 3.10	Curve of in situ pore water pressure at a depth of 11.5 m	55
Fig. 3.11	Curve of over-static pore water pressure and vibration number. (a) 8.5 m. (b) 11.5 m. (c) 13.5 m	56
Fig. 3.12	Comparison between measured and fitted curves. (a) 8.5 m. (b) 11.5 m. (c) 13.5 m	57
Fig. 3.13	Curves of increase rate of pore water pressure	58
Fig. 3.14	Deformation of soft clay	65
Fig. 3.15	Curves of dynamic strain and vibration number under different CSR	66
Fig. 3.16	Curves of dynamic strain and vibration number at different cyclic stress ratio. (a) CSR = 0.1. (b) CSR = 0.2. (c) CSR = 0.3. (d) CSR = 0.4	67
Fig. 3.17	Curves of critical cyclic stress ratio verse vibration number	68
Fig. 3.18	Soil deformations at different consolidation state. (a) $K_c = 1$ and $N = 1{,}000$. (b) $K_c = 0.46$ and $N = 1{,}000$	69
Fig. 3.19	Curves of dynamic strain and confined pressure in different vibration number	70
Fig. 3.20	The natural logarithm relation of dynamic responses and loading time: (a) dynamic stress; (b) dynamic shear strength	72
Fig. 3.21	Step cyclic loading during the cyclic triaxial test	76
Fig. 3.22	Typical sinusoidal cyclic loading	77
Fig. 3.23	Initial cyclic creep characteristic curve during the typical stage	77
Fig. 3.24	Mechanical characteristics of the soil sample in the initial five cycles: (a) creep strain; (b) pore water pressure; (c) axial stress strain	79
Fig. 3.25	Mechanical characteristics of the soil sample in the last five cycles: (a) creep strain; (b) pore water pressure; (c) axial stress strain	80
Fig. 3.26	Recoverable elastic strain versus cyclic stress amplitude curves with loading frequency: (a) $f = 0.5$ Hz; (b) $f = 2.5$ Hz	82
Fig. 3.27	Fitting lines between recoverable elastic strain and cyclic stress ratio with loading frequency: (a) 0.5 Hz; (b) 2.5 Hz	82
Fig. 3.28	Loading frequency versus recoverable elastic strain under different cyclic stress ratios	83
Fig. 3.29	Accumulated plastic strain versus number of cycles with loading frequency: (a) $f = 0.5$ Hz; (b) $f = 2.5$ Hz	84
Fig. 3.30	Comparative curves of accumulated plastic strains with different loading frequencies (a) $f = 0.5$ Hz; (b) $f = 2.5$ Hz	85
Fig. 3.31	Residual pore water pressure versus number of cycles with loading frequency: (a) $f = 0.5$ Hz; (b) $f = 2.5$ Hz	86

Fig. 3.32	Curves dynamic stress and strain of soft silt clay	88
Fig. 3.33	Curves of $1/E_d - \varepsilon_d$	88
Fig. 3.34	Curves of $E_d - \varepsilon_d$	89
Fig. 3.35	Curves of $1/G_d \sim \gamma_d$	89
Fig. 3.36	Curves of $\tau_d \sim \gamma_d$	90
Fig. 3.37	Working principles of GDS BES	91
Fig. 3.38	Working interface of GDS BES	91
Fig. 3.39	Deformation of specimen after test	92
Fig. 3.40	Curves of effective principal stress versus vibration cycles number	93
Fig. 3.41	Curves of shear wave velocity versus time	96
Fig. 4.1	Cross section through the center of the pore channels (pore throat d and ventral pore d')	102
Fig. 4.2	Microstructure of undisturbed samples under different multiples (**a**) 638×; (**b**) 5,557×; (**c**) 6,887×	110
Fig. 4.3	Microstructure of samples after vibration under different multiples (**a**) 580×; (**b**) 2,319×	113
Fig. 4.4	X-ray diffraction curve of clay minerals: (**a**) Air drying; (**b**) EG saturation; (**c**) heating treatment (cps is the unit of diffraction intensity. It refers to the electron number received by counter per seconds)	116
Fig. 4.5	Microstructure of clay skeleton	117
Fig. 4.6	Microstructure of samples after vibration (×5,000): (**a**) horizontal surface of undisturbed sample; (**b**) horizontal surface of sample after vibration; (**c**) vertical surface of undisturbed sample; (**d**) vertical surface of sample after vibration	118
Fig. 4.7	Arrangement of mineral particles: (**a**) face-to-face; (**b**) edge-to-edge; (**c**) edge-to-face	118
Fig. 4.8	Contact state of particles: (**a**) vertical surface and (**b**) horizontal surface	119
Fig. 4.9	SEM images of saturated soft clay soil in Shanghai: (**a**) salt crystals; (**b**) aerial pore structure; (**c**) biological fabrics	120
Fig. 4.10	Classification of pore structure in clay soil	121
Fig. 4.11	SEM picture of undisturbed sample	122
Fig. 4.12	SEM picture of NO. 830D1 sample after vibration	123
Fig. 4.13	SEM picture of NO. 826D2 sample after vibration	123
Fig. 4.14	SEM picture of NO. 828D3 sample after vibration	124
Fig. 4.15	Dynamic shear stress–strain curve	125
Fig. 4.16	Automatic mercury analyzer AUTOSCAN 60	128
Fig. 4.17	Sample of mercury analyzer	128
Fig. 4.18	Analysis system for low pressure and high pressure: (**a**) low pressure; (**b**) high pressure	129
Fig. 4.19	Curve of pore diameter versus pressure of mercury intrusion	132

Fig. 4.20	Curve of volume versus pressure of mercury intrusion	133
Fig. 4.21	Curve of specific surface area of accumulation versus pressure of mercury intrusion	133
Fig. 4.22	Distribution of accumulation pores	134
Fig. 4.23	Curve of differential pore volume distribution	134
Fig. 4.24	Curve of $D_v(r)$ with pore distribution	135
Fig. 4.25	Curve of $D_s(r)$ with pore distribution	135
Fig. 4.26	Curve of volume percent versus pressure of mercury intrusion	136
Fig. 4.27	Curve of volume of intrusion (extrusion) versus pressure of mercury intrusion: (**a**) intrusion curve; (**b**) extrusion curve	137
Fig. 4.28	Distribution of pore diameters	137
Fig. 4.29	Fractal dimension of undisturbed sample calculated through Menger model	143
Fig. 4.30	Fractal dimension of undisturbed sample 830D1 calculated through Menger model	143
Fig. 4.31	Fractal dimension of undisturbed sample 826D2 calculated through Menger model	144
Fig. 4.32	Fractal dimension of undisturbed sample 828D3 calculated through Menger model	144
Fig. 4.33	Fractal dimension of undisturbed sample calculated through the thermodynamics model	145
Fig. 4.34	Fractal dimension of sample 830D1 calculated through the thermodynamics model	146
Fig. 4.35	Fractal dimension of sample 826D2 calculated through the thermodynamics model	146
Fig. 4.36	Fractal dimension of sample 828D3 calculated through the thermodynamics model	147
Fig. 4.37	Relationship between fractal dimension D_T and CSR	148
Fig. 4.38	Relationship between cyclic stress ratio and mean distribution radius	148
Fig. 4.39	Relationship between cyclic stress ratio and retention factor total porosity	149
Fig. 4.40	Relationship between cyclic stress ratio and specific surface area	149
Fig. 4.41	Relationship of cyclic stress ratio and retention factor	150
Fig. 4.42	Relationship of mean distribution radius and axial strain	150
Fig. 4.43	Relationship of total porosity and axial strain	150
Fig. 4.44	Relationship of retention factor and axial strain	151
Fig. 4.45	Relationship of specific surface and axial strain	151
Fig. 5.1	Simplified model for the interaction in a wheel-rail system	167
Fig. 5.2	(**a**) The motion of the wheel with flat under low speed and (**b**) the motion of the wheel with flat under high speed	170
Fig. 5.3	The motion of an eccentric wheel	171

List of Figures

Fig. 5.4	Schematic diagram of a subway train	173
Fig. 5.5	Schematic diagram of subway train load	174
Fig. 5.6	Stratigraphic profile of the soil model	177
Fig. 5.7	Strata generation	178
Fig. 5.8	Strata interfaces	178
Fig. 5.9	Tunnel lining and rail	179
Fig. 5.10	Geometric model	179
Fig. 5.11	Example of a meshed finite element model	180
Fig. 5.12	Diagram of mesh generation	183
Fig. 5.13	Typical wave curve of soil response obtained through numerical simulation	187
Fig. 5.14	Contour plot of response amplitude of horizontal stress in cross section	188
Fig. 5.15	Contour plot of response amplitude of horizontal stress in longitudinal section	189
Fig. 6.1	Longitudinal settlement curve along the tunnel axis	193
Fig. 6.2	Gaussian normal distribution of settling tank	195
Fig. 6.3	Filed measured curve of settling tank	195
Fig. 6.4	Theoretical curve of grouting filling rate and land subsidence	196
Fig. 6.5	Curves of grouting filling rate and measured settlement	197
Fig. 6.6	Curve of gap coefficient and grouting volume in practice	199
Fig. 6.7	The settlement curve at B5 monitoring point of No. 2 shield (lasting 50 days)	199
Fig. 6.8	Accumulated settlement of Line 1, Shanghai Metro from 1995 to 1999 (After Chen and Zhan 2000)	204
Fig. 6.9	Cumulative plastic strains of soils in the literature	206
Fig. 6.10	Application of NNGM (1, 1) in laboratory tests	211
Fig. 6.11	Monitoring data of tunnel settlement of Metro Line 4 from Helen Road Station to Baoshan Road Station in Shanghai (Wang 2011; Ren et al. 2012)	212
Fig. 6.12	Comparisons of GM (1, 1) and NNGM (1, 1)	216
Fig. 6.13	Comparison of NNGM (1, 1) and NEM	217
Fig. 6.14	Relative errors of NNGM (1, 1) and NGM (1, 1)	218
Fig. 6.15	Ground settlement curve from March 2000 to December 2002	220
Fig. 6.16	Comparison chart of fitting curve and measured curve	221

List of Tables

Table 2.1	Frequency statistics	31
Table 2.2	Fitting parameters	33
Table 3.1	Scheme of undrained dynamic triaxial test	46
Table 3.2	Basic physical mechanic index of the 4th layer of saturated soft silt clay	46
Table 3.3	Model parameters	55
Table 3.4	Values of fixed parameters C	57
Table 3.5	Attenuation model parameters of pore water pressure	59
Table 3.6	Basic physical mechanic index of test soil	63
Table 3.7	Dynamic strength index at $K_c = 1$, $\varepsilon_p = 5\%$	72
Table 3.8	Control parameters of cyclic triaxial tests	75
Table 3.9	Relevant parameter values	89
Table 3.10	Variation parameters of effective principal stress	94
Table 3.11	Parameters of decline curves of shear wave velocity	97
Table 4.1	Characteristic parameters of pore structure of undisturbed and vibrated samples	138
Table 4.2	Characteristic parameters of pore structure of soil samples before and after vibration	139
Table 4.3	Fractal dimension calculated through Menger model	145
Table 4.4	Fractal dimension calculated through the model of thermodynamics correlation	147
Table 5.1	Depth of the stratums	177
Table 5.2	Results of wave velocity field test	181
Table 5.3	Physical and mechanics parameters of soil	184
Table 5.4	Material parameters of subway lining and rail	184
Table 5.5	Comparison of response amplitude of stress between calculation and field test along the depth at 1.8 m from the tunnel edge	188

Table 5.6	Comparison of response amplitude of stress between calculation and field test along the horizon at 13.5 m depth	188
Table 6.1	Relationship between measured settlement and grouting quantity along the axis	197
Table 6.2	Comparison between calculated value and measured value	198
Table 6.3	Reference level of model accuracy (Liu et al. 2010)	210
Table 6.4	Accuracy class of NNGM (1, 1)	212
Table 6.5	Reference properties of subgrade soils of Metro Line No. 4 in Helen Road Station	213
Table 6.6	Predicted results of tunnel settlement based on in-site monitoring data	214

Brief Introduction for the Book

This book is written based on the research project "Micro-structure Changes and Dynamic Characteristics of Saturated Soft Clay Around the Subway Tunnels Under Train Loading" which is mainly supported by Natural Science Foundation of China (Grant No. 40372124 and No. 41072204). The primary study object of this book is the dynamic loading induced by subway trains and dynamic responses of the soft clay around tunnels. Microscopic and macroscopic methods were applied in the experimental study and theoretical analysis. Field test was conducted to study the characteristics of the dynamic load induced by the subway train; it was also applied to obtain the distribution law of the dynamic stress in the soft clay around the tunnel. According to the analysis of the field monitoring data, a series of laboratory tests were conducted to study the dynamic response of the soft clay under different dynamic loading (frequency, amplitude, and loading times). Meanwhile, we conducted corresponding microscopic tests for further understanding of the deformation mechanism of soft clay under dynamic loading. Through all the above tests, the dynamic strain–stress relationship, the generation and dissipation of excess pore water pressure, and the changes of the microstructure were discussed in the book. This book also discussed the ground settlement induced by subway train loading, and its long-term prediction was studied as well. The conclusions obtained from this study are meaningful to the theoretical development and practical construction of subway system in soft clay area.

This book can be used as a textbook for graduate students of geological engineering, geotechnical engineering, tunneling engineering, and disaster prevention engineering. Meanwhile, it can be applied as a reference for technical staffs and managers in the practical construction.

Chapter 1
Introduction

Recently, subway has become the major part of urban transportation. Rapid development of economy in China has spurred and accelerated the urban traffic congestion problem. With lots of people migrating into large and medium cities, the previous existing urban transportation hardly meets the demand of the population expansion and urban sprawl. The subway system brings free-flow traffic at all times and realizes multidimensional space traffic to make faster modernization of urban development. At present, many large modern cities have interlaced the heartland by subways, such as Shanghai, Nanjing, and Shenzhen. A large number of medium cities are constructing subway systems. Even some small cities are applying for subway construction to alleviate the traffic jam. Hereby, the subway construction has a really great prospective.

Saturated soft clay areas (including Shanghai, Tianjin, Hangzhou, and Guangzhou) are facing several severe issues of high population, narrow space, crowd transportation, etc. Establishing spacing transportation network is the key to alleviate the situation of traffic congestion, in which subway plays a significant role to develop the traffic underground space without any large disturbance to the surface construction. Subway is favored by urban residents as a new type transportation vehicle with characteristics of high-speed, safe, comfortable, low-energy consumption and pollution. It plays a great role in improving resident's living environment, increasing the urban land utilization, easing the city traffic, shunting various transport modes in stereo, and preserving city historical and cultural landscapes. The advantages are highly recognized all over the world.

As people's living standards improve, the vibration effects on environment have been paid much attention now. In Shanghai, a stereo space traffic system has been gradually developed and mostly is concentrated in dense resident area and high-rise commercial centers in the city. Under the continuous vehicle loading, large differential settlement will occur in soft soil foundation even after long-term consolidation. In the transportation system planning to the next five-year, totally 200-km subway lines will be constructed. And the long-term plan is to invest

200 billion into subway construction, which will make 780 km mileage in total. Most subways are located in or through the saturated soft clay soil layer based on the subways on operation and in construction. Some segments of tunnel have over 20 cm settlement in axis. It affected the normal operation of subways according to the monitoring data in tunnel and surrounding environment. Moreover some old house near the subway showed up cracks. All these are related to the microstructure damage under subway loading. The occurrence and development is an evolutionary process from quantitative to qualitative.

Soft clay soil is a typical fine soil with natural void ratio over 1.0 and natural water content greater than liquid limit. It mainly includes sludge, muddy soil, peat, and cumulosol. Since soft clay soil is characterized by large void ratio, high water content, poor permeability, high compressibility, and low shear strength, all these properties result in unique and special engineering behaviors. The influence factors of the strength and deformation of soft clay are very complicated. Therefore, systematical and deep study on dynamic properties, micro-deformation, and damage mechanism can effectively control the subway axis deformation, avoid or mitigate the land subsidence, provide reference for subway construction design, and predict the subway settlement as well. As for the physical and mechanical response, stress-strain properties, microstructure damage characteristics, and process of soft clay, it can provide significant theoretical and practical meanings. And finally it is helpful to provide reference to establish the prediction and prevention system of urban environmental geological disaster.

Hereby, this monograph presents the study on dynamic response characteristics of mucky clay under subway vehicle loading, based on two National Natural Science Foundation projects "study on dynamic properties and corresponding micro-structure influences of soft clay soil around subway tunnel under vibration loading (Grant No. 40372124)" and "study on the engineering properties and deformation characteristics of saturated clay surrounding subway tunnel after artificial ground freezing in soft clay area (Grant No. 41072204)," combined the field monitoring data of pore pressure and soil stress and strain with laboratory tests, including unidirectional excitation CKC cyclic triaxial tests, GDS dynamic cyclic triaxial test, scanning electron microscopy (SEM), and automatic mercury intrusion porosimetry (MIP). Comprehensive analysis with micro-macro, static-dynamic methodologies, the microstructure deformation, and failure of soft clay under subway vibration loading is studied in depth via collecting response frequencies, soil strain, and pore pressure variation subjected to subway loading. The pore water pressure development and consolidation deformation characteristics of soft clay soil under subway traffic loading are summarized here, and the comparison of results in computational numerical simulation with experiments is discussed to analyze the subway vibration loading action and influence on soft clay soil. Finally, the microstructure variation properties are quantitatively analyzed to reveal the deformation and damage mechanism.

1.1 Research Status

1.1.1 The Study on Soil Structure

Soil structure herein means the properties of soil particles and pores, including their distribution, connection status, and modes. The structure behavior especially refers to mechanical effects of the soil structure, i.e., the mechanical response of the soil structure when subjected to loadings. Generally speaking, the natural normal consolidated soil, over-consolidated soil, or unconsolidated soil each has unique structural properties. Heterogeneity and structural properties are the main characteristics differing soil from the other engineering materials. Most natural sedimentary soils have structural behaviors in different extent. Usually only those with visible structural effects are named structural soils, in particular obvious mechanical effects of the soil particle arrangement and connection.

The structural properties have significant influence on soil mechanical behaviors. Firstly, structural soils have visible structural yield stress on the compression curve. The compression properties differ a lot after yield stress. The pseudo-over-consolidated soil induced by structural behaviors has much discrepancy compared with caused by stress histories. At the same stress condition, pores of the former are larger. Secondly, the stress-strain relationship greatly relates to the stress level. When the consolidation stress is lower than the structural yield stress, the soil behaves as stress softening. Conversely, it acts as stress hardening, when the consolidation stress is larger than structural yield stress. In the undrained consolidation shear test, the pore water variation develops with the difference between consolidation stress and structural yield stress. It is possible that the coefficient of Skempton may be greater than 1. Thirdly, the coefficients of consolidation and permeability vary a lot as soil structure changes. Before the natural soil structure is damaged, the coefficient of consolidation is almost a constant. But it depresses sharply once the soil structure is disturbed. The permeability of undisturbed soil is larger than remolded soil by 5–10 times. At last, strength envelop of the structural soil behaves as a fold line, in which clear turning point occurs at the structural yield strength.

Therefore, the structural behavior is an important characteristic that should be paid much attention in soft soils. It plays a significant role on physical and mechanical properties, particularly the compression behavior under dynamic loadings. Meantime, the soil structure study provides theoretical support for the micro-deformation research on soils.

1.1.2 The Study on Vehicle Vibration Loading

To study the dynamic response and properties of saturated soft clay under subway vibration loading, basically we should figure out the main characteristics of the subway vibration loading. Limited by the methodologies and apparatus, the research on vehicle vibration loading does not proceed for a long period all over the world. Since the 1970s, the influences of vehicle vibration on surrounding environment and constructions have been gradually paid attention. Some results are achieved as follows.

Pan and Pande (1984) and Liang and Cai (1999) utilized some kind of excitation load function to express the vehicle loading, which can reflect vehicle vibration loading simply, namely,

$$F(t) = A_0 + A_1 \sin \varpi_1 t + A_2 \sin \varpi_2 t + A_3 \sin \varpi_3 t, \qquad (1.1)$$

where $F(t)$ is the train vibration load, A_0 is the vehicle static load, and A_1, A_2, A_3 are the amplitudes of some frequency, respectively. When the subway train advances at a constant velocity V, rail vibration wave length L_i and corresponding amplitude α_i are measured as well, then all the angular velocity can be obtained corresponding to each L_i by $\varpi_i = 2\pi V/L_i$, and then each vibration amplitude can be calculated out as $A_i = m \cdot \alpha_i \cdot \varpi_i^2$.

Gan and Cao (1990) analyzed the random process simulation of vertical vibration on bridge by virtue of bringing the track irregularity function into dynamic equilibrium equations. They simulated a specific spectral density function of a railway line in China, which was provided by the Central South University Railway College, to get that variable function of the track irregularity needed in simulation. Li and Li (1995) assumed the train had an infinite length and considered each wheel set as a periodic moving load addressed by Fourier series. They converted the additional dynamic loading induced by vertical vibration of train into an exponential function of vibration amplitude and frequency. Accounting for all wheel loadings of crabbing as well, the superposition of wave was applied, which was suitable for wave propagation elements. Luo and Geng (1999) established a wheel-rail contact model by introducing rail irregularity based on random vibration theory. They assumed there was rigid linear contact stiffness between wheels and rail to get the motion equation and deduced mean square deviation of dynamic loading of the lower part of spring by Fourier transform. As for the upper part of spring, the upper half of trains were subjected to the train vibration motion equations and the mean square deviation of dynamic loading was obtained by Fourier transform and superposition methodology. Nelson (2000) discussed the Biot wave propagation model in isotropic porous medium by seismic reflection method and combined the multi-degree of freedom railway vehicle operation mode. He predicted the ground vibration motion in heterogeneous isotropic saturated soil or rock excited by railway vehicle operation. It was found that saturation acted to decay the vibration response greatly. And in addition to the vertical force of vehicles, the bumping of wheels was an important source of ground vibration. Gao et al. (2003) determined the vehicle vibration loading by a certain wheel-rail model. They obtained the system vibration mode and frequency through modal analysis of subway structure system to determine reasonable damping coefficient and time step, which were used in Newmark implicit time integration method. The dynamic response of short-distance overlap tunnel in segment was analyzed under three cases of inbound-line dynamic load, outbound-line dynamic load, or intersection load. Finally weak parts of the lining structure and corresponding displacement and stress were figured out under vehicle vibration load. Sheng et al. (2003a, b) established a theoretical ground vibration model, consisting of two kinds of loading mechanisms, quasi-static and dynamic. Coupling all the operation dynamics of a series of vehicles moving along an infinite railway track, the influence of vibration frequency was analyzed when train velocity was above or below the minimum wave velocity of ground vibration. Through the comparison of measured data and model analysis in three places, results indicated dynamic load can generate higher level vibration. Paolucci et al. (2003) analyzed and simulated the ground vibration induced by train operation. Spectral element discrete method was used to establish 2D and 3D models. Rail vibration and attenuation of the peak ground wave velocity with distance were both discussed in detail. Degrande et al. (2006) proposed a numerical model to predict the vibration generated by subway operation in tunnel. Tunnel and soil were fitted in different way in this model and the boundary misconvergence was limited. Comparison was made in the dynamic responses of two different types of tunnel inverts under harmonic; one was a Paris subway tunnel buried in a shallow sand layer and another was a London subway in a deep clay layer.

Besides above theoretical computations and numerical analysis methods, field testing is the best important and direct way to determine the subway vibration loads. There is also a lot of relevant experimental research work by scholars all over the world.

Some field vibration tests were conducted both in 1977 at Berlin, Germany, and in 1982 at London, England. Subsequently, Japanese scholar Takemiya (2003) also tested the vibration caused by Ledsgard high-speed subway operation. He confirmed his field testing results combined numerical simulation. Relevant research in China started relatively later. Spurred by rapid development in urban subway, the environmental problems induced by subway vibration are paid more attention gradually. Researches in this area emerge much more.

Pan and Xie (1990) and Pan et al. (1995) performed a field test of subway vibration at the section of Chongwenmen-Qianmen in Line 2, Beijing Subway. Some relevant numerical simulation was conducted as well. Zhang and Bai (2000) derived the rail vibration acceleration formula responding to subway train vibration through spectral analysis method based on field testing data. They simplified vehicle system vibration to establish a kinematic equation of rail system and then back-inferred the subway vibration load. Wang et al. (2005) utilized an advanced instrument to test dynamic response of the tunnel structure at Zhuting area in a railway line from Beijing to Guangzhou. They figured out the response frequency properties by analyzing the field data and then established train vibration load model. Zha et al. (2007) did a field test on a low embankment construction section of Lianyan highway. They used strong motion seismograph and dynamic earth pressure transducers to detect the dynamic response for different vehicle types, different vehicle velocities, different depths of embankment, and ground motion as well. Li et al. (2007) detected the vibration generated by Shanghai Maglev train under different velocities. The ground motion propagation properties were analyzed. Jin et al. (2005) derived out vibration equations based on kinematic equilibrium equation solution by Newmark method. They proposed a dynamic finite element method to back calculate the vibration load duration from vibration acceleration. Liang et al. (2006) opined that vehicle vibration load was a complex problem, involving train axle load, suspension systems, vehicle speed, rail composition and track regularity, etc. Considering the mechanism of subway vibration, the superposition of wheel-rail contact forces between adjacent wheel set, stress dispersion by sleepers, vibration excited by track irregularity, etc. were all taken into account for the modification of the existed empirical equations of vehicle vibration load.

1.1.3 Dynamic Response of Vehicle Vibration Load

At present, dynamic responses of underground structure systems to vehicle vibration load are concentrated in two aspects. First is the environmental influence; second is the dynamic response of tunnel structure. As for the environmental vibration corresponding to vibration load, previous research was mainly about the empirical equations by mathematically processing from field testing data. In the 1970s, Lang (1971) proposed a simple prediction formula to reveal the relationship of vibration level and distance. Subsequently, Kurzweil (1979) provided different correction parameters for various train types, tracks, tunnels, and buildings to predict the environmental vibration in different locations. In China, Wang (1993) took environmental testing survey on subway operation in Beijing and Tianjin and analyzed the line environment circumstance along subways. Gu et al. (1996) studied the subway vibration propagation path and influence parameters. A simple quantitative prediction formula for subway vibration was summarized and confirmed by field testing results. Cui and Ji (1996) predicted the influence of Nanjing Subway on ground environment by empirical equations as well.

After the 1980s, the scholars all over the world have been involved in further advanced research on subway vibration. Numerical simulation, laboratory tests, and combined field testing methods are all used to predict dynamic responses of underground structures to subway vibration loads. In Germany, Rucker (1980) applied two-dimensional finite element method to simulate and discuss the vibration propagation variations at different depth along tunnel during subway operation. Eisenmann et al. (1983) provided minimum distance of buildings from subway tunnel corresponding to various installation system of different rail type. Nelson et al. (1983) comprehensively and systematically summarized all environmental impact factors of the subway system in North America, supported by the Massachusetts Department of Transportation. Metrikine and Vrouwenvelder (2000) considered a tunnel structure as an infinitely extended Euler beam buried in clay layers and then simply analyzed the ground motion response properties of foundation in two dimensions subjected to a moving load. Sheng et al. (2003a, b) studied the vibration of layered-soil foundation in which the circular tunnel was exerted to simple harmonic vibration load. Forrest and Hunt (2006a, b) brought interface units into the consideration of interaction of tunnel structure and foundation by utilizing cylindrical shell and infinitely radial extended elastic soil. Dynamic responses of the circular tunnel were analyzed when subjected to normal point load. Andersena and Jones (2006) comparatively analyzed dynamic response in 2D and 3D by boundary element method. Cloutean et al. (2005) used substructure method to discuss the coupling dynamic behavior of tunnel and foundation. In China, Pan and Xie (1990) got the dynamic response at several feature points in some tunnel section of a Beijing subway by virtue of field testing data and analyze them by spectral analysis method. Then they derived a simulation mathematical formula for train vibration load according to the track acceleration testing data. System dynamic properties were figured out through numerical simulation in tunnel and surrounding soils. Zhang and Pan (1993) also gave a train vibration load function based on field testing data. And they applied plastic constitutive relationship and Mohr-Column yield criterion in a FEM-IEM coupling model to conduct the dynamic analysis in time domain. Li (1997) simulated the vertical vibration load through field testing data and analyzed the dynamic response in tunnel lining by FEM. The safety and stability of tunnel lining structure were evaluated under vertical vibration load. Chen et al. (1998) arranged a large number of testing points along subway track in Weishanzhuang-Andingdaokou section of Line 2 in Beijing Subway. They mathematically processed 108 acceleration wave curves generated by 18 subway train operation and figured out a correlation function in which mean acceleration decayed with distance. One profile was chosen to calculate the static force distribution, natural frequency, dynamic response, and the relations between the dynamic response and the structure influence range. Gao (1998; Gao et al. 2003) analyzed the dynamic properties of tunnel and surrounding soils in the tunnel of Beijing-Tongliao Line by FEM. Wang and Xia (1999) used a simplified method to establish a two-dimensional model of intersection analysis on subgrade soil building. The train vibration wave propagation properties and influence on adjacent buildings were calculated. Zhang and Xia (2001) simulated and analyzed the dynamic response of platform and surrounding buildings in Bawangfen area (Sihui Station in Batong Line, Beijing Subway). Displacement responses in time domain under train vibration load were calculated in a number of points on buildings. Hereby, the variation of response vibration level was analyzed under several different circumstances, including various moving speeds of train, different building structure parameters, different floors on the same building, and the buildings with different distance along subway axis. Zhang (2001) derived the subway vibration load function by field testing data in some tunnel section in Shanghai subway. Accordingly he analyzed the dynamic response under that vibration load function in short-distance overlap tunnel in subway section between Nanpu Bridge and South Pudong Road in Phase II project of Line 3 in Shanghai Metro. The evaluation of mechanical behaviors was conducted on tunnel structure. Gong et al. (2001, 2004) did the laboratory on dynamic pore water pressure variation in foundation soil, taking long-term vibration load with some interval on a single subway train into account. They determined the critical dynamic stress and critical dynamic strain in the foundation soil of Line 1, Nanjing Metro. The results under different loads indicated that the dynamic pore water pressure model could be divided into fatigue type and enhancement type. Finally the dynamic pore water pressure in Nanjing foundation soils was regarded as enhancement type according to the dynamic stress calculation in foundation soils, and corresponding dynamic pore water pressure computation model was established. Xie et al. (2002) simplified the rail-foundation system as a Winkle beam model on a semi-infinite layered-soil foundation and analyzed the synergistic action between beam and foundation under high-speed moving load. Accordingly the dynamic responses of rail and ground surface were figured out. Xie et al. (2004) discussed the foundation deformation problem under moving loads and deduced deformation computation equations under various loading modes. They took a triangular load as an example to compare the influences of several different factors on foundation deformation, including track, natural frequency, moving speed, and foundation type. Wang and Chen (2005) studied the stress properties and distribution to find that the response vibration generated by subway traffic load actually was a cyclic load mainly on compression stress. The dynamic stress distribution was similar to that generated by static load (such as strip load) on foundation surface but dynamically changed with subway train moving. In addition, they also studied the influence of soil type in semi-infinite foundation and train speed on stress. It was found that the influence of shear modulus on stress related to the train speed. The train speed greatly affected the stress distribution, while Poisson's ratio just had some impact on lateral stress. Li et al. (2005) used excitation function to simulate vertical vibration load of high-speed train. They analyzed the dynamic response of large cross-section tunnel structure under train vibration load in detail. The influence of cross-section type, train speed, and damping ratio on vibration responses was mainly focused. Wang et al. (2006) based on the subway vibration field testing results in Zhuting tunnel in Beijing-Guangzhou line analyzed the dynamic response properties of tunnel lining structure under three different cross sections. The vertical displacement, vertical acceleration, and all kinds of internal stress time duration curves of tunnel lining structure were obtained. The research results were favorable for the evaluation of dynamic stability of tunnel lining structure and provided same reference for perfecting the theoretical design of railway tunnel structure. Bian (2006a, b) combined finite element method and thin-layer element method to establish 2.5-dimensional model, analyzing the three-dimensional problem on foundation under moving load of close or over critical speed. The three-dimensional problem was converted into a plain stress condition with 3 degrees of freedom in each point by wave-number transform along load moving direction. The calculation efficiency was improved greatly. Finite element method was utilized for simulating the area close to foundation within the loading range, while dynamic thin-layer element method to establish transmission boundary to simulate vibration wave propagation from near field to infinite area. Zhou et al. (2006) studied the dynamic response of pore water pressure in saturated clay under vibration load, based on continuous monitoring of pore water pressure in different location and along different depth near some segment in Line 2 of Shanghai Metro. The variations of pore water pressure dissipation and increase in saturated

1.1 Research Status

clay were figured out, and some explanation was inferred by soil dynamics and energy conservation principle. Bian (2006a, b) proposed a quasi-analytical method to analyze the dynamic response of viaducts under high-speed train vibration load. He mainly considered the dynamic interaction of bridge and foundation and divided the problem into two models by substructure method: First was 3D FEM dynamic model of the viaduct under train vibration load and second was the intersection effect model of piles and soils, in which the axis-symmetrical group pile foundation based on Fourier series expansion coordinated the surrounding layered soils with the continuity conditions in nodes on pile foundation cap. The influence of far field foundation on near FEM area was simulated by establishing transmission boundary conditions of the stress wave in thin-layer element method. The impedance function was used to represent the supporting act of group file foundation to upper bridge structure. Through numerical simulation, the dynamic properties of the Shinkansen viaduct on soft foundation under high-speed vibration loads were investigated. The influences of axle load, moving speed, and group pile foundation on viaduct vibration properties were analyzed accordingly. He et al. (2007a, b) derived the coupling vibration equation of rail and foundation according to the solving theory of wave equation of layered foundation soil in wave-number domain: A unified general expression in wave-number domain was acquired. The viscous damping was used in study material. The energy transfer characteristics and influence of simple harmonic vibration load on vibration decaying were both discussed by virtue of solving the vibration differential equations in frequency-wave number domain and then applying inverse Fourier transform to get ground motion response. A simulation analysis on ground vibration generated by train axle load was performed. In addition, He et al. (2007a, b) established a 3D numerical simulation model on ground vibration under moving axle load based on soil dynamics and finite element method. The vibration propagation characteristics in ground soil and the damping effect of isolation trench were analyzed under three different train moving speed. Wang et al. (2009) compared dynamic response monitoring data of the train operating vibration in Jinliwen railway and a rock blasting vibration on K17 slope treatment in 330 State Road nearby the railway tunnel. Results showed that the dynamic response to blasting vibration was a gradually decaying curve with a single vibration source; while in the train vibration, it behaved as coupling attenuation by a series of consecutive vibration source. Chen (2007) conducted a pulsation testing, a train excitation test, and a train excitation frequency test in the tunnel section of People's Square-Xinzha Road of Line 1, Shanghai Metro. The field response duration curves were obtained. Zhang (2007) preset earth pressure and pore water pressure transducers along depth at various locations around tunnel section of Jing'an Temple-Jiangsu Road in Line 2, Shanghai Metro, and performed a continuous dynamic monitoring in field. Response frequency of saturated soft clay, the soil response stress amplitude variation with distance perpendicular to subway tunnel, and also the variation along the depth were all discussed. The dynamic decaying formula was proposed, and the influence range and corresponding amplitude value were derived accordingly. Li et al. (2008) applied high-sensitivity acceleration transducers and dynamic data collector and recorder to the field testing at the Dongdan-Jianguomen tunnel section in Batong Line, Beijing Subway. The vertical vibration acceleration generated by subway operation and surface traffic was acquired. The ground vibration properties induced by subway operation were studied and the testing processing was conducted on ground motion response to subway vibration.

1.1.4 Dynamic Properties of Soft Clay Soil Under Dynamic Loading

Subway traffic load refers to a kind of long-term cyclic load. Due to its special loading form and the ubiquity with subway, study on dynamic properties of soils under cyclic load has great theoretical meaning and practical significance. Since the 1960s, a lot of scholars have been involved in this area and many valuable research results have been achieved all over the world. The study is mainly concentrated on the methods of laboratory tests and computational numerical simulation, etc. The influences of cyclic stress ratio, number of cycles, vibration frequency, over-consolidation ratio, and drainage condition during loading are comprehensively considered on the dynamic properties of soils. Variations of accumulated plastic deformation and residual pore water pressure are mainly acquired. Outside China, the main achievements have been made as follows:

Seed and Chan (1961) did a dynamic strength test of saturated soft clay. They noticed that additional deformation occurred under the dynamic load in a deformation-stable specimen after consolidation. And they pointed out that keeping the dynamic stress frequency and duration the same, this additional deformation amount can be determined by consolidation pressure, dynamic stress, and number of loading cycles. Lo (1969) investigated the pore water pressure under cyclic load and defined a term named pore water pressure ratio related to vertical pre-consolidation stress. He discussed the relation between pore water pressure ratio and strain and opined that loading time, consolidation pressure, and time had no influences on this special relation. Luo (1973) systematically studied the residual deformation of clay under cyclic load and established an empirical equation of relation between loading cycles N and residual deformation ε_p to predict the vibration induced settlement in earth dam.

$$\varepsilon_p = 10\left\{\frac{\left[C_4' + S_4'\ (K_c - 1) + C_5'\sigma_3 + S_5'\ (K_c - 1)\sigma_3\right]^{-1}\sigma_d}{(0.1N)^{S_1}}\right\}^{C_6+S_6'(K_c-1)} \quad (1.2)$$

where K_c was consolidation ratio, σ_d was dynamic stress, and C_4', S_4', C_5', S_5', C_6', S_6', and S_1 were test parameters, varying with soil properties. Matsui et al. (1980) carried out cyclic triaxial tests on remolded normally consolidated and over-consolidated clay specimens, in which the mean primary stress was kept as a constant. The pore water pressure was measured precisely during the tests, and factors influencing pore

1.1 Research Status

water pressure were analyzed. Subsequently, post-undrained and drained triaxial tests were conducted, and the relationships between strength, stiffness, and equivalent over-consolidation ratio were qualitatively analyzed. Yasuhara et al. (1982) performed several groups of cyclic stress-controlled triaxial tests on remolded saturated soft clay. The results indicated that undrained cyclic strength was almost independent on the frequency and duration of cyclic load; cyclic strength was a little lower than static strength due to remolding in samples; normalized pore water pressure u/σ'_3 had a hyperbolic relationship with cyclic shear strain. Yasuhara (1985) also predicted the deterioration properties of remolded high-plasticity clay under various confining pressures, loading frequencies, and cyclic stress levels. Results showed the effective stress path of apparent over-consolidated status induced by undrained cyclic load was much similar to variation of effective stress generated by stress release in over-consolidated clay. Hereby, he derived the calculation formula to predict the strength variation based on apparent over-consolidation ratio. Fujiwara et al. (1985, 1987) comprehensively discussed factors such as total load, loading period, load increment ratio, loading form, degree of cementation, and number of loading cycles influencing the deformation on clay. It was found that consolidation deformation under cyclic load was larger than that generated in static load, in which he elucidated the reason was secondary consolidation that occurred when the cyclic load was applied. And the corresponding amount in secondary consolidation was directly proportional to cyclic load value, load increment ratio, loading period, and organic content. Hyodo et al. (1992, 1999) processed the results of undrained triaxial tests on high-plasticity marine clay and introduced two parameters of relative cyclic strength and effective stress spatial location into empirical model to simulate the pore water pressure generation and dissipation. Yilmaz et al. (2004) collected mucky clay around the M7.4 strong earthquake location in Tokyo, Japan, and performed standard triaxial and cyclic triaxial tests, respectively, on samples. Results showed that the dynamic strength and stiffness were rarely decayed by the laboratory load equivalent to the strong earthquake. And the accumulated strain was greatly dependent on loading regime and loading stress level.

In China research in this subject started relatively late. Initial study focused on accumulated deformation and pore water pressure variation under cyclic load. Later the research was much concerned about the prediction model and computation method on these two aspects. As for the strength and stiffness variation of soils subjected to cyclic load, it has been rarely found. Yang (1990) analyzed the constitutive relationship of soil under cyclic load and predicted the deformation during loading. He proposed a computation mode for residual deformation under multistage undrained cyclic load based on laboratory cyclic triaxial tests, in which drained full consolidation of specimen was permitted and pre-shearing action was underscored. Yan (1991) discussed the deformation properties of remolded soft clay and proposed a concept of equivalent static load. He tried to simplify complex dynamic loads into an equivalent static load and then predicted the deformation of soil. Zhang and Tao (1994) carried out cyclic triaxial tests on saturated soft clay with 21.6 plasticity index I_P. Results implied that under a certain number of loading cycles, an increase in loading frequency can result in a rise of the dynamic pore water pressure. Zhou et al. (1996) gave a computation model of residual strain and pore water pressure in soft clay under dynamic load based on laboratory testing data. The validation of this model was made as well to ensure its applicability by comparing the computation results to the measured pore water pressure and residual strain in undrained and partially drained conditions. Finally by means of a case study, the settlement induced by dynamic loads was calculated and analyzed. He (1997) found that the static loading rate and dynamic loading frequency had no obvious influence on stress or strain in sands but as for clay influence existed according to test results. When loading frequency was lower than 5 Hz, the dynamic modulus and damping ratio increased greatly with ramping frequency. Zhou et al. (2000) carried out stress-controlled cyclic triaxial test and analyzed the influence of cyclic stress ratio, loading cycles, loading frequency, over-consolidation ratio on strain, and pore water pressure in normally consolidated saturated soil in Hangzhou. The critical cyclic stress ratio and threshold cyclic stress ratio were determined. Jiang and Chen (2001) discussed the consolidation characteristics of clay under cyclic loads in several different wave shapes. It was found that there were three critical points in consolidation deformation, which divided the whole consolidation process into four stages. The previous three stages were akin to behaviors under static load, while the last stage was unique in cyclic load. Usually the phenomenon of piping or boils occurs during this stage and the deformation generated exceeded the amount under the static load. In addition wave shape of cyclic load had no influence on the position of critical points. Tang et al. (2003, 2004) studied the dynamic strain development of muddy silty clay under subway cyclic load. Some factors of consolidation status, consolidation ratio, axial cyclic stress, and frequency were fully considered. The variations of critical cyclic stress ratio and dynamic strain were figured out under different loading cycles, loading frequencies, confining pressures, and consolidation status. Moreover, the dynamic shear modulus and dynamic shear strength were acquired by dynamic stress-strain relationship. Chen et al. (2006) summarized a series of laboratory tests of shear velocity compression tests and cyclic triaxial tests on undisturbed and remolded soil samples and concluded that under static loads, deformation of undisturbed soils had three stages along the stress level, which were divided by pre-consolidation pressure and structure yield stress. While in a remolded soil sample, the deformation was rarely related with stress level. Under the cyclic load, these two kinds of soils both have a turning point in the curve of strain and numbers of cycles under different consolidation pressure. This turning point changed with dynamic load amplitude. But the strain was still linearly related to cycle number at failure. Huang et al. (2006) carried out undrained cyclic triaxial tests on typical saturated soft clay in Shanghai. They analyzed several main factors influencing accumulated plastic deformation of soft clay: loading cycles, initial deviator stress, and dynamic deviator stress. According to theories in critical soil mechanics, he brought a relative deviator stress level parameter into analysis on undrained accumulated deformation of saturated soft clay under different combination loading mode of static and cyclic stress conditions. Wang et al. (2007a, b) studied softening effects of stiffness and strain of normally consolidated saturated soft clay load by triaxial cyclic loading tests in

Xiaoshan, Hangzhou. An empirical formula was derived to reflect the development of stiffness softening, in which a failure stiffness ratio of soft clay was attained. He also proposed an empirical equation for strain softening of clay. Combined with modified Iwan model, a dynamic stress-strain relationship of soft clay was depicted. Chen et al. (2005) performed a series of undrained cyclic triaxial tests on undisturbed, remolded and cement mixed clay samples in Xiaoshan, Hangzhou. Based on the comparison results of undisturbed and remolded samples, undrained instantaneous accumulated deformation properties of soft clay were studied under cyclic load with and without static deviator stress. They proposed a constitutive model of soil with several factors considered, including cyclic stress, number of loading cycles, over-consolidation ratio, and static deviator stress. Results indicated that the deformation properties of soft clay were greatly influenced by consolidation stress level and structure damage degree. When the consolidation pressure was larger than structure yield stress, the deformation property was close to the behavior in remolded soils. There was a turning point in the testing sample under cyclic loads, which implied the failure of soil structure. The constitutive model should be described respectively before or after the sample structure damaged.

1.1.5 Microstructure of Soft Clay Soil

Researchers all over the world study the soil structure mainly from three aspects: macroscopic, mesoscale, and microscale levels. The macrostructure refers to natural or undisturbed visible soil structural properties. The meso-structure means the structural properties that can be observed from thin section or optical section by polarizing microscopy, in which the structure unit can be 0.05–2 mm, i.e., the aggregates consisting of sand, silt, native mineral particles, and clay. Structural properties in microscale can be obtained with a variety of modern techniques of scanning electron microscopy (SEM) or X-ray diffractometry (XRD). Through knowledge of microstructural characteristics, the nature of many engineering behaviors can be understood better. Particularly, it provides significant basis for establishing applicable constitutive relationship and makes reasonable sense to some macro-properties.

Throughout the history of study on microstructure in clay all over the world, it boomed in the 1960s and 1970s; a large number of publications emerged. In 1973, the first international microstructure conference was held in Switzerland. There were a lot of papers in the fifth international conference on soil mechanics. From the late 1970s, the research rush gradually slowed down and achievements on microstructure were rarely published out. While at the same time, some great progress of computer quantitatively processing on microstructure morphology was attained in clayey soils, by some research groups led by Tovey and Osipov, et al. Some new breakthrough was made out by the Bazant's research group in establishing micro soil mechanism in clay.

In China, relevant research work was started relatively decades later. Until the late 1970s, the specific microstructure study was actually carried out. With the popularity of scanning electron microscopy (SEM) in China, many scholars were involved in this special issue. It culminated with large amount of papers in the 1980s. The research work revealed some characteristics during this time (Shi 1996a, b; Shi et al. 1997; Shi and Jiang 2001):

1. High starting point. Even though the research work in China started almost decades later than other European countries, the onset of study began at a very high level, based on international high technology. Many researchers utilized the world's most advanced electron microscopy testing method to conduct their work and stood on the shoulders of giants.
2. Rare in pure analysis of microstructure morphology. It was mainly concentrated in qualitative study on engineering properties combined with clayey soil (especially loess, expansive soil, soft soil, etc.). There was great progress in some explanation or inference of special engineering properties by virtue of soil structure.
3. The microstructure sampling technology has greatly improved. Li and Wang (1985) and Wu (1988) successfully developed a frozen vacuum sublimation dryer, respectively. It makes great breakthrough on the microstructure sampling technology in China.
4. The quantitative analysis of microstructure has started. Tan et al. (1980) and Shi et al. (1988) quantitatively determined the flat clay mineral particle orientation in clayey soil by X-ray diffractometer. Shi and Li (1995) quantitatively analyzed the microstructure morphology by computer image processing technology. Great progress was achieved.

The main success and achievement in microstructure are as follows:

Shi (1996a, b) introduced some new technology, new results, and new progress in microstructure aspect of engineering clayey soil, mainly including freeze drying method for high-water content sample in SEM analysis, and quantitatively evaluated the image processing technology in microstructure of clayey soil, EDX technology to determine the chemical composition of inter-aggregate cement in clayey soil, and CT technology for no damage detection for soil structure. Gong (2002) analyzed the particle and aggregate composition, pore size distribution (PSD), microstructure, and cation exchange capacity with pore solution. He discussed the variation of PSD after consolidation and analyzed the potential influence of artificial recharge on soil properties. The significant role of microstructure playing on soil consolidation and surface subsidence was explained from physical and chemical aspects.

Wang et al. (2003) confirmed the quantitative analysis method on microstructure properties of cement mixed soil by processing microstructure images of soft soil after cement reinforcement. He proposed a microstructure properties evaluation method by characterization of structural factors such as structural unit, pore size, shape, and orientation. Li and Bao (2003) utilized the grayscale in microstructure images of soil to calculate the porosity. He proposed an algorithm to obtain porosity based on the grayscale values in three-dimensional spaces. Wang et al. (2004) deter-

1.1 Research Status

mined the threshold value used in image processing for microstructure analysis by combining image processing technology with GIS software to get fractal dimensions of particle morphology of the soil. He demonstrated the fractal characteristics of clayey soil by analyzing the microstructure images of the expansive soil in Yun Town in China. The relation between fractal dimension and soil microstructure was derived by processing the microstructure images.

Liang et al. (2005) developed a new information extraction technology by combining GIS software MapInfo with image processing software Photoshop. He proposed an M-P method to extract all kinds of data from SEM images for analyzing the microstructure of clayey soils.

Tang et al. (2005) analyzed the microstructure characteristics of saturated soft clay by SEM. He further studied the microstructure unit shape and size, contact state, connection, and pore shape and size. The deformation mechanism was elucidated from microstructure view by comparing sample subjected to triaxial cyclic load with those undisturbed soils. Tang et al. (2007) quantitatively analyzed the microstructure of saturated soft clay surrounding the subway tunnel by mercury intrusion porosimetry, based on the field monitoring and laboratory triaxial cyclic tests. The results showed that the pore structure fractal dimension variation at different depths can be reflected by different cyclic stress ratio (CSR).

Wang et al. (2007a, b) respectively analyzed the two-dimensional and three-dimensional porosity algorithms by processing SEM images based on area and volume calculation method by GIS.

Tang et al. (2008a) calculated the apparent porosity and particle morphological fractal dimensions by a large series of SEM images to determine the optimum threshold value. Influence factors such as threshold value, size of analysis area, scanning position, and magnification on microstructure of soils were discussed. The mechanism of each influence was explored and best threshold value and magnification were recommended. Zhou and Lin (2005) acquired the relation between some microstructure characterization parameters and soil strength through direct shear tests and triaxial consolidation tests of undisturbed soft soil samples of marine deposits from central region in Pearl River Delta. And they further explored the macro-micro connection. Zhou and Lin (2008) subsequently started from laboratory consolidation tests and studied the microstructure properties variations and consolidation deformation characteristics. They proposed a settlement prediction model of soft soil by extracting microstructure parameters and establishing the relation between those parameters with deformation and strength. Finally this model was applied into a practical engineering project, and the calculation value was really close to monitoring results.

Chen and Wang (2008) reinforced various organic-content soft soils with cements and analyzed the relation between cement strength and organic content by unconfined compression tests. The microstructure images from SEM were quantitatively analyzed by image processing system WD-5. The particle morphological fractal dimension values were derived to deploy the geometry fractal dimension of particle size distribution. The results indicated that the increase in organic content resulted in larger fractal dimension of particle size distribution, smaller degree of aggregates, and poor the cement reinforcement effect. The organic content had some influence on the effectiveness of cement reinforcement on soft clay. The fractal dimension of particle size distribution was negatively correlated with the strength of cement-reinforced soft soil. Zhou et al. (2009) utilized the image processing software Image Pro Plus to analyze the microstructure images of the clay minerals in deep soft rock. Pore structure information was extracted and fractal dimension was numerically analyzed. Results implied that the uniformity of pore size distribution had an inverse-proportional relationship with fractal dimension.

1.1.6 The Long-Term Settlement of Soft Clay Foundation Under Vehicle Vibration Loading

Post-construction settlement and differential settlement occurred in the soft foundation when subjected to long-term cyclic loads, which greatly affects the foundation's performance and service life. With the increase in engineering practice, the prediction of long-term settlement in soft foundation is paid more attention. Presently, the foundation settlement calculation is concentrated in construction deformation and seismic settlement, and post-construction long-term settlement under cyclic loads is rarely mentioned.

Currently the method for calculating accumulated cyclic settlement mainly involves two aspects. One is the layered summation based on practical experience. Second is the dynamic analysis method based on elastoplastic constitutive model, in which the dynamic consolidation equation and dynamic elastoplastic constitutive model are utilized; the accumulated deformation and pore water pressure properties are described using elastoplastic constitutive model under cyclic load by finite element method; the settlement and pore water pressure duration developments of soft foundation are calculated under cyclic loading step by step. This dynamic analysis method is much accordant with the mechanism of soil deformation and pore water pressure dissipation. But the difficulty is that when the loading cycles reach tens of thousands or hundreds of thousands, a huge amount of calculation is generated and it's marginally harder for application. Li et al. (2006) preliminarily analyzed the accumulated settlement of the soft foundation under cyclic loads using isotropic elastoplastic bounding surface model and dynamic consolidation theory. But limited by the computation time, the cyclic times was not consistent with the actual situation. Relatively speaking, the first method is simple formulized and the parameters can be easily determined. It is widely used in engineering practices. As for the research on this aspect, the theoretical and experimental studies both started relatively late. At present, more scholars have combined theory and experiment to develop the applicable method. Many researchers introduced Terzaghi quasi consolidation theory to calculate the consolidation settlement due to pore water dissipation.

1.1 Research Status

Hyodo et al. (1996) predicted the foundation deformation under traffic loads by two-dimensional numerical dynamic analysis, based on the laboratory results of dynamic triaxial tests. But the practical foundation deformation is three-dimensional. This method could not simulate the engineering practice specifically. Kutara et al. (1980) employed one-dimensional consolidation theory to predict the long-term settlement under traffic loads, which is equivalent to the static load. But the three-dimensional properties and settlement accumulation effect were not considered in their computation. Ling et al. (2002) calculated the accumulated settlement and pore water pressure using empirical equations. And then the accumulated settlement of the subgrade was derived by superposition in layered summation under traffic loads. Current research work indicated that the physical meaning of the computation equation used in engineering practice was not explicit and the influence of stress-strain history on accumulated settlement could not be considered.

Chai and Miura (2002) performed the additional settlement calculation generated on the road of Saga Airport in Japan under traffic loads, without considering the accumulated pore water pressure dissipation during the cyclic loading. Li et al. (2006) determined the undrained shear strength by effective consolidation stress method, based on undrained accumulated settlement calculation proposed by Chai and Miura (2002). They applied this method into the additional settlement prediction of the low subgrade after construction in Shanghai outer loop line road. Even though this method can consider the initial static stress, the consolidation deformation generated by the dissipation of accumulative pore water pressure under traffic loading could be taken into account. Hence, Li et al. (2006) proposed a cyclic accumulated deformation and pore water pressure model based on the critical state theory. The influence of static deviator stress and dynamic stress level interaction can be rationally considered. He predicted the long-term settlement of subgrade in Shanghai outer loop line road by this model and then calculated the settlement induced by undrained cyclic accumulated deformation and pore water pressure dissipation by layered summation method, which was added up into the total accumulated settlement.

Liu (2006) developed a new method by combining the quasi-static finite element computation and empirical fitting calculation model to predict the long-term settlement in saturated soft clay foundation. The total settlement was divided into two parts: One was the accumulated deformation induced by undrained cyclic loading, and second was the consolidation deformation during the accumulated pore water pressure dissipation by dynamic loading. These two parts were calculated by empirical fitting formulas respectively based on critical state theory. The stress components were extracted from the finite element computational results. Finally, the total settlement was attained by superposition in layered summation.

Tang et al. (2008b) determined the dynamic elastic module of saturated soft clay by laboratory triaxial cyclic tests and then utilized it into the numerical simulation for calculating the deformation of soils surrounding subway under vibration loads. The final ground settlement was computed further under the subway vibration load.

Bian and Chen (2006) establish a three-dimensional analysis model for the dynamic interaction of track and foundation under vehicle moving loads, by combining 2.5-dimensional finite element method with thin-layer element method. The dynamic deviator stress distributions in the underlying foundation layers were derived under various train speed. The long-term dynamic additional settlement computation method was proposed based on the accumulated plastic deformation theory under the cyclic load.

Wei and Huang (2009) acquired an empirical model of residual deformation development of soft soil under long-term reciprocating loads, based on previous experimental results. According to the quasi-static method, the traffic load was considered as a long-term cyclic distributed static load. The additional dynamic stress in foundation was calculated by elastic layered theory. And the proposed model was used to attain the residual strain in each layer. The final settlement was obtained by integrating the residual strain along depth.

Wei et al. (2008) established the long-term settlement prediction model of subway tunnel using ant colony algorithm, in which comprehensively considering all settlement influence factors, information functions and heuristic functions were built by the longitudinal accumulated settlement data and accumulated differential settlement data, based on the field monitoring data of settlement along some sections of Line 1, Shanghai Metro.

1.2 Research Content and Methodology

The pore water pressure models differed due to different sample preparations (undisturbed or remolded), different initial consolidation conditions (isotropic consolidation or anisotropic consolidation), different loading methods (constant stress controlled or constant strain controlled), etc. There is no unified model in this aspect. And the previous research was concentrated in the cyclic triaxial tests under high stress level condition (cyclic stress ratio is greater than 0.05). The study on pore water pressure development in saturated soft clay under subway traffic loads is rarely involved.

As for the deformation and damage of saturated soft clay under subway vibration loading, it firstly behaved as pore water pressure accumulation and dissipation; at the same time, the microstructure changed, such as particle rearrangement and porosity variation. Despite there already exists a lot of achievement on the microstructure study of soil, the specific research on microstructure variation of saturated soft clay under subway traffic loading can be rarely found.

Therefore, this monograph firstly acquired parameters in fielding monitoring tests for the dynamic response analysis of saturated soft clay to subway vibration loading. Secondly, laboratory tests were conducted, including unidirectional excitation CKC cyclic triaxial tests and GDS dynamic cyclic triaxial test, to summarize the pore water pressure characteristics and consolidation deformation properties under subway vibration loads. Then, the influence on surrounding soil by subway vibration loads was discussed and performed by numerical simulation. Finally, the microstructure variations of mucky clays when subjected to subway vibration loads were quantitatively analyzed by scanning electron microscopy (SEM) and automatic mercury intrusion porosimetry (MIP). The damage mechanism was further explored.

1.2.1 Dynamic Response of Soft Clay Surrounding Subway Tunnel to Subway Traffic Loads

The field monitoring tests were conducted near Jing'an Temple Station of Line 2, Shang Metro. Boreholes were drilled for the installation of pore water pressure and earth pressure transducers along various depths at different distances from the subway tunnel. The earth pressure variations and pore water pressure developments were all monitoring at real time. Comparison on different locations and depths was performed to figure out the relations between earth pressure and pore water pressure characteristics under subway traffic loads and horizontal distance or vertical depth.

1.2.2 Dynamic Characteristics of Soft Clay Under Subway Vibration Loads

This chapter presents all laboratory results based on the parameters acquired from field tests, including unidirectional excitation CKC cyclic triaxial tests and GDS dynamic cyclic triaxial test. The dynamic properties of pore water pressure, soil dynamic strength, dynamic stress-strain relation, dynamic elastic module with vibration frequency, and cyclic time were all further discussed. The microstructure deformation and damage mechanism were inferred accordingly by scanning electron microscopy (SEM) and automatic mercury intrusion porosimetry (MIP).

1.2.3 Microstructure Study of Soft Clay Under Subway Vibration Loads

This part quantitatively and qualitatively analyzed the microstructure variation of soft clay after subjected to subway vibration loads using SEM and MIP methods. The variation of pore size distributions, porosity, and other pore structure parameters were derived by MIP, and pore structure characterization was described by SEM images. Hereby, the relation of macro-mechanical behavior was explored based on the microstructure parameters.

1.2.4 Numerical Simulation

Numerical simulation model was established based on the field and laboratory tests. The dynamic response properties of soils around subway tunnel were simulated by a three-dimensional finite element model. The land subsidence induced by long-term settlement under subway vibration loads was computationally deduced in this section.

1.2.5 Settlement Prediction of Soil Surrounding Subway Tunnel Under Subway Vibration Loads

Based on the previous research results in deformation mechanism of soft clay, we comprehensively analyzed all kinds of settlement influence factors and predicted the settlement by an unequal interval GM (N, 1) model. In this model, all influence factors were considered by using Newton quadratic polynomial interpolation method to transform the unequal interval sequence into equal interval sequence. The predicted results were compared with field monitoring data to confirm the model's validation.

References

Andersena L, Jones CJC (2006) Coupled boundary and finite element analysis of vibration from railway tunnels-a comparison of two and three dimensional models. J Sound Vib 293(611):625
Bian XC (2006a) Analysis of viaduct-ground vibrations due to high-speed train moving loads. J Vib Eng 19(4):438–445 (in Chinese)
Bian XC (2006b) Ground vibration due to moving load at critical velocity. J Zhejiang Univ (Eng Sci) 40(4):672–675 (in Chinese)
Bian XC, Chen YM (2006) Ground vibration generated by train moving loadings using 2.5D finite element method. Chin J Rock Mech Eng 25(11):2335–2342 (in Chinese)
Chai JC, Miura N (2002) Traffic-load- induced permanent deformation of road on soft subsoil. J Geotech Geoenviron Eng 10:907–916 (in Chinese)
Chen CX (2007) Spot dynamics experiment and analysis of operated metro tunnel in soft soil. Dissertation, Tongji University, Shanghai, China
Chen HE, Wang Q (2008) Fractal study on microstructure of cement consolidated soil. J Harbin Inst Technol 40(2):307–309 (in Chinese)
Chen S, Xu GB, Gao R (1998) Dynamic analysis of the influence of train railway vibration on the building along railways. J North Jiaotong Univ 22(4):57–60 (in Chinese)
Chen YP, Huang B, Chen YM (2005) Deformation and strength of structural soft clay under cyclic loading. Chin J Geotech Eng 27(9):1065–1071 (in Chinese)
Chen YM, Chen YP, Huang B (2006) Experimental study of influence of static and cyclic deformation on structural soft clay with stress level. Chin J Rock Mech Eng 25(5):937–945 (in Chinese)
Cloutean D, Arnst M, Al-Hussaiui TM (2005) Freefield vibrations due to dynamic loading on a tunnel embedded in a stratified medium. J Sound Vib 283(1-2):173–199

References

Cui ZX, Ji ZY (1996) Prediction of the impact of subway vibration on the ground environment. Noise Vib Control 1:9–14 (in Chinese)

Degrande G, Clouteau D, Othman R, Arnst M, Chebli H, Klein R, Chatterjee P, Janssens B (2006) A numerical model for ground-borne vibrations from underground railway traffic based on a periodic finite element-boundary element formulation. J Sound Vib 293:645–666

Forrest JA, Hunt HEM (2006a) A three-dimensional tunnel model for calculation of train-induced ground vibration. J Sound Vib 294(678):705

Forrest JA, Hunt HEM (2006b) Ground vibration generated by trains in underground tunnels. J Sound Vib 294:706–736

Fujiwara H, Yamanouchi T, Yasuhara K (1985) Consolidation of alluvial clay under repeated loading. Soil Found 25(3):19–30

Fujiwara H, Shunji U, Kazuya Y (1987) Secondary compression of clay under repeated loading. Soil Found 27(2):21–30

Gan HL, Cao XQ (1990) Analysis of the influence of increasing the train speeds on the vertical vibration of bridge through stochastic simulation. J Shanghai Railw Univ 11(4):45–53 (in Chinese)

Gao F (1998) Analysis of dynamic responses of a railway tunnel subjected to train loading. J Lanzhou Railw Inst 17(2):6–12 (in Chinese)

Gao F, Guan BS, Qiu WG, Wang MN, Li CH (2003) Dynamic responses of overlapping tunnels to passing trains. J Southwest Jiaotong Univ 38(1):38–42 (in Chinese)

Gong SL (2002) The microscopic characteristics of Shanghai soft clay and its effect on soil mass deformation and land subsidence. J Eng Geol 10(4):378–384 (in Chinese)

Gong QM, Liao CF (2001) Testing study of dynamic pore water pressure under train loading. Chin J Rock Mech Eng 20:1154–1157 (in Chinese)

Gong QM, Zhou SH, Wang BL (2004) Variation of pore pressure and liquefaction of soil in metro. Chin J Geotech Eng 26(2):290–299 (in Chinese)

Gu XA, Liu XZ, Zhang CH (1996) The method study on preduction of subway environment vibration. Environ Eng 14(5):35–39

He CR (1997) Dynamic triaxial test on modulus and damping. Chin J Geotech Eng 19(2):39–48 (in Chinese)

He ZX, Zhai WM, Yang XW, Chen XW (2007a) Moving train axle-load induced ground vibration and mitigation. J Railw Sci Eng 4(5):73–77 (in Chinese)

He ZX, Zhai WM, Yang XW, Chen XW (2007b) Semi-analytical study of the ground vibration under axial loading. J Vib Shock 26(12):1–7 (in Chinese)

Huang MS, Li JJ, Li XZ (2006) Cumulative deformation behavior of soft clay in cyclic undrained tests. Chin J Geotech Eng 28(7):891–895 (in Chinese)

Hyodo M, Yasuhara K, Hirao K (1992) Prediction of clay behavior in undrained and partially drained cyclic tests. Soil Found 32(4):117–127

Hyodo M, Yasuhara K, Murata H (1996) Deformation analysis of the soft clay foundation of low embankment road under traffic loading. In: Proceeding of the 31st symposium of Japanese Society of Soil Mechanics and Foundation Engineering, pp 27–32

Hyodo M, Hyde AFL, Yamamoto Y, Fujii T (1999) Cyclic shear strength of undisturbed and remoulded marine clays. Soil Found 39(2):45–48

Jiang J, Chen LZ (2001) One-dimensional settlement due to long-term cyclic loading. Chin J Geotech Eng 23(3):366–369 (in Chinese)

Jin LX, Zhang JS, Nie ZH (2005) Dynamic inverse analysis for vibration-load history of high-speed railway. J Traffic Transp Eng 5(1):36–38 (in Chinese)

Kurzweil LG (1979) Ground-borne noise and vibration from underground rail systems. J Sound Vib 66(3):363–370

Kutara K, Miki H, Seki K, Mashita Y (1980) Settlement and countermeasures of the road with low embankment on soft ground. Tech Rep Civil Eng JSCE 22(8):13–16

Lang J (1971) Results of measurements on the control of structure-borne noise from subways. In: Seventh International Congress on Acoustics, Akadémiai Kiadó, Budapest, pp 421–424

Li DW (1997) An analysis of dynamic response of tunnel lining to train vibrations. J Lanzhou Railw Inst 16(4):24–27 (in Chinese)

Li Q, Bao SS (2003) The computation of the porosity of saturated clay based on the gray scale of soil microstructure graph. J Chongqing Normal Univ Nat Sci Ed 20(1):30–31 (in Chinese)

Li JS, Li KC (1995) Finite element analysis for dynamic response of roadbed of high-speed railway. J China Railw Soc 17(1):66–75 (in Chinese)

Li SL, Wang ZH (1985) The characteristics of Chinese fine soil distribution on plasticity chart. Yantu Gongcheng Xuebao 7(3):84–89

Li L, Zhang BQ, Yang XL (2005) Analysis of dynamic response of large cross-section tunnel under vibrating load induced by high speed train. Chin J Rock Mech Eng 24(23):4259–4265 (in Chinese)

Li JJ, Huang MS, Wang YD (2006) Analysis of cumulative plastic deformation of soft clay foundation under traffic loading. China J Highw Transp 19(1):1–5 (in Chinese)

Li W, Gao GY, Sun YM (2007) Test analysis of ground vibration caused by maglev train in Shanghai. Shanxi Arch 33(24):269–271 (in Chinese)

Li WD, Liu WN, Zhang HR (2008) Test analysis of metro induced ground vibrations at interval. China Railw Sci 29(1):1–7

Liang B, Cai Y (1999) Dynamic analysis on subgrade of high speed railways in geometric irregular condition. J China Railw Soc 21(2):84–88 (in Chinese)

Liang SH, Sun RH, Li WP (2005) Using Mapinfo and Photoshop to study SEM images of clay. J Henan Univ Sci Technol Nat Sci 26(1):55–58

Liang B, Luo H, Sun CX (2006) Simulated study on vibration load of high speed railway. J China Railw Soc 28(4):89–95 (in Chinese)

Ling JM, Wang W, Bu HB (2002) On residual deformation of saturated clay subgrade under vehicle load. J Tongji Univ 30(11):1315–1320 (in Chinese)

Liu M (2006) Cyclic constitutive modeling of soft clays and long-term predictions of subway traffic-load-induced settlements. Dissertation, Tongji University, Shanghai, China

Lo KY (1969) The pore pressure-strain relationship of normally consolidated undisturbed clays. Can Geotech J 6:383–412

Luo WK (1973) The characteristics of soils subjected to repeated loads and their applications to engineering practice. Soil Found 13(1):11–27

Luo YY, Geng CZ (1999) Influence of track state on vertical wheel/track dynamic overloads. J China Railw Soc 21(2):42–45 (in Chinese)

Matsui T, Ito T, Ohara H (1980) Cyclic stress-strain history and shear characteristics of clay. J Geotech Eng 106(10):1101–1120

Metrikine AV, Vrouwenvelder ACWM (2000) Surface ground vibration due to a moving train in a tunnel: two-dimensional model. J Sound Vib 234(1):43–66

Nelson GT (2000) Prediction of ground vibration from trains using seismic reflectivity methods for a porous soil. J Sound Vib 231(3):727–737

Pan SC, Pande GN (1984) Preliminary deterministic finite study on a tunnel driven in loess subjected to train loading. China Civil Eng J 17(4):19–28 (in Chinese)

Pan CS, Xie ZG (1990) Measurement and analysis of vibrations caused by passing trains in subway running tunnel. China Civil Eng J 23(2):21–28 (in Chinese)

Pan CS, Li DW, Xie ZG (1995) The study of the environmental effect of the vibration induced by Beijing subway. J Vib Shock 14(4):29–34 (in Chinese)

Paolucci R, Maffeis A, Scandella L, Stupazzini M, Vanini M (2003) Numerical prediction of low-frequency ground vibrations induced by high-speed trains at Ledsgaard Sweden. Soil Dyn Earthq Eng 23:425–433

Seed HB, Chan CK (1961) Effect of duration of stress application on soil deformation under repeated loading. In: Proceedings 5th international Congress on soil mechanics and foundations, Dunod, Paris, pp 341–345

Sheng X, Jones CJC, Thompson DJ (2003a) A comparison of a theoretical model for quasi-statically and dynamically induced environmental vibration from trains with measurements. J Sound Vib 267:621–635

References

Sheng X, Jones CJC, Thompson DJ (2003b) Ground vibration generated by a harmonic load moving in a circular tunnel in a layered ground. J Low Freq Noise Vib Act Control 22(2):83–96

Shi B (1996a) Quantitative assessment of changes of microstructure for clayey soil in the process of compaction. Chin J Geotech Eng 18(4):57–62 (in Chinese)

Shi B (1996b) Review and prospect on the microstructure of clayey soil. J Eng Geol 4(1):39–44 (in Chinese)

Shi B, Jiang HT (2001) Research on the analysis techniques for clayey soil microstructure. Chin J Rock Mech Eng 20(6):864–870 (in Chinese)

Shi B, Li LS (1995) The quantatively analysis of SEM image on the microstucture of clay. Chin Sci (Special A) 25(6):666–672

Shi B, Wang BJ, Ning WW (1997) Micromechanical model on creep of anisotropic clay. Chin J Geotech Eng 19(3):7–13 (in Chinese)

Shi et al (1988) The relationship between the microstructure and the engineering properties of compacted expansive soil

Takemiya H (2003) Simulation of track-ground vibrations due to a high-speed train: the case of X-2000 at Ledsgard. J Sound Vib 261:503–526

Tan et al (1980) The anisotropy behaviors of clay on the mesostructure. Hydrol Geol 5:4

Tang YQ, Huang Y, Ye WM, Wang YL (2003) Critical dynamic stress ratio and dynamic strain analysis of soils around the tunnel subway train loading. Chin J Rock Mech Eng 22(9):1566–1570 (in Chinese)

Tang YQ, Zhang X, Ye WM, Huang Y, Wang YL, Zhou NQ (2004) A study on dynamic strength and dynamic stress-strain relation of silt soil under traffic loading [J]. J Eng Geol (S1):98–101

Tang YQ, Zhang X, Zhou NQ, Huang Y (2005) Microscopic study of saturated soft clay's behavior under cyclic loading. J Tongji Univ Nat Sci 33(5):626–630 (in Chinese)

Tang YQ, Zhang X, Zhao SK (2007) A study on the fractals of saturated soft clay surrounding subway tunnels under dynamic loads. China Civil Eng J 40(11):86–91

Tang CS, Shi B, Wang BJ (2008a) Factors affecting analysis of soil microstructure using SEM. Chin J Geotech Eng 30(4):560–565 (in Chinese)

Tang YQ, Luan CQ, Zhang X, Wang JX, Yang P (2008b) Numerical simulation of saturated soft clay's deformation around tunnel under subway vibration loading. Chin J Undergr Space Eng 4(1):105–110 (in Chinese)

Wang XX (1993) Prediction of the impact of subway vibration on the environment along the subway line. Noise Vib Control 5:22–24 (in Chinese)

Wang CJ, Chen YM (2005) Analysis of stresses in train-induced ground. Chin J Rock Mech Eng 24(7):1179–1186 (in Chinese)

Wang FC, Xia H (1999) Vibration effects of trains on surrounding environments and buildings. J North Jiaotong Univ 23(4):13–17 (in Chinese)

Wang Q, Chen HE, Cai KY (2003) Quantitative evaluation of microstructure features of soil contained some cement. Rock Soil Mech 24:12–17 (in Chinese)

Wang BJ, Shi B, Tang CS (2004) Fractal study on microstructure of clayey soil by GIS. Chin J Geotech Eng 26(2):244–247 (in Chinese)

Wang XL, Yang LD, Gao WH (2005) In-situ vibration measurement and load simulation of the raising speed train in railway tunnel. J Vib Shock 24(3):99–102

Wang XQ, Yang LD, Zhou ZG (2006) Dynamic response analysis of lining structure for tunnel under vibration loads of train. Chin J Rock Mech Eng 25(7):1337–1342 (in Chinese)

Wang BJ, Shi B, Tang CS (2007a) Study on 3D fractal dimension of clayey soil by use of GIS. Chin J Geotech Eng 29(2):309–312 (in Chinese)

Wang J, Cai YQ, Xu CJ, Liu W (2007b) Study on strain softening model of saturated soft clay under cyclic loading. Chin J Rock Mech Eng 26(8):1713–1719 (in Chinese)

Wang H, Liu DL, Chen JJ (2009) Comparative study of dynamic response of tunnels under the vibrated load induced by blast and subway train. Railw Eng 12:69–72 (in Chinese)

Wei X, Huang MS (2009) A simple method to predict traffic-load-induced permanent settlement of road on soft subsoil. Rock Soil Mech 30(11):3342–3346 (in Chinese)

Wei K, Gong QM, Zhou SH (2008) Forecast of long-term settlement of metro tunnel on the basis of ant colony optimization. J China Railw Soc 30(4):79–83 (in Chinese)

Wu YX (1988) The quantitively analysis of microstructure of engineering clay [D]. Graduate School of Chinese Academy of Geological Sciences, Beijing

Xie WP, Hu JW, Xu J (2002) Dynamic response of track-ground systems under high speed moving load. Chin J Rock Mech Eng 21(7):1075–1078 (in Chinese)

Xie WP, Wang GB, Yu YL (2004) Calculation of soil deformation induced by moving load. Chin J Geotech Eng 26(3):318–322 (in Chinese)

Yan PW (1991) The deformation characteristics of remolded soft clay under repeated loading. Chin J Geotech Eng 13(1):48–53 (in Chinese)

Yang QJ (1990) Constitutive relationship and deformation prediction of soft soil under cyclic loading. Dissertation, Wuhan University of Hydraulic and Electrical Engineering, Wuhan, China

Yasuhara K (1985) Undrained and drained cyclic triaxial tests on a marine clay. Proc 11th ICSMFE, Rotterdam 2:1095–1098

Yasuhara K, Yamanouchi T, Hirao K (1982) Cyclic strength and deformation of normally consolidated clay. Soil Found 22(3):77–79

Yilmaz MT, Pekcan O, Bakir BS (2004) Undrained cyclic shear and deformation behavior of silt-clay mixtures of Adapazari in Turkey. Soil Dyn Earthq Eng 24:497–507

Zha WH, Hong BN, Xu Y (2007) Vibration measurement and analysis of low embankment freeway pavement and roadbed under Traffic Loadings. Highw Eng 32(4):113–117 (in Chinese)

Zhang P (2001) Dynamic response analysis of close overlapping subway tunnels under subway train load. Dissertation, Tongji University, Shanghai, China

Zhang X (2007) The study on the micro-structure and dynamic characteristic of soft clay around tunnel under the subway-induced loading. Dissertation, Tongji University, Shanghai, China

Zhang YE, Bai BH (2000) The simulation of the vibration load on the tunnel induced by subway train. J Vib Shock 19(3):68–78 (in Chinese)

Zhang YE, Pan CS (1993) Tests and analysis of the dynamic response of tunnels subjected to passing train load. J Shijiazhuang Railw Inst 6(2):7–14 (in Chinese)

Zhang KL, Tao ZY (1994) The prediction of pore pressure of saturated clay under cyclic loading. Rock Soil Mech 15(3):9–17 (in Chinese)

Zhang N, Xia H (2001) Impact of the subway vibration on the buildings adjacent to the subway. Eng Mech 3:199–203

Zhou CY, Lin CX (2005) Relationship between micro-structural characters of fracture surface and strength of soft clay. Chin J Geotech Eng 27(10):1136–1141 (in Chinese)

Zhou CY, Lin CX (2008) Research on deformation calculation model of soft soil based on microstructure. ACTA Sci Nat Univ Sunyatseni 47(1):16–20 (in Chinese)

Zhou J, Tu HQ, Yaswhara K (1996) A model for predicting the cyclic behavior of soft clay. Rock Soil Mech 17(1):54–60 (in Chinese)

Zhou J, Gong XN, Li JQ (2000) Experimental study of saturated soft clay under cyclic loading. Ind Constr 30(17):43–47 (in Chinese)

Zhou NQ, Tang YQ, Wang JX (2006) Response characteristics of pore pressure in saturated soft clay to the metro vibration loading. Chin J Geotech Eng 28(4):2149–2152 (in Chinese)

Zhou L, Du WX, Han X (2009) Fractal characteristics of micro pore structure for clay mineral. J Heilongjiang Inst Sci Technol 19(2):94–97 (in Chinese)

Chapter 2
Field Tests

2.1 Introduction

Being safe, comfortable, and fast, subway becomes essential in urban transportation. The vibration problem induced by subway operation also should be paid attention. There are a variety of models to analyze the ground vibration caused by the subway loading (Alabi 1992; Lipen and Chigarev 1998; Sheng et al. 1999; Jones et al. 2000). Some papers (Pan et al. 1995; Hirokazu et al. 2001; Xie et al. 2002) studied the dynamic response of track system under subway traffic loading, which is concentrated in the superstructure, and foundation soils are rarely involved. Currently, dynamic responses of the saturated soft clay around the subway tunnel have not been studied yet at home and abroad, while greater deformation of saturated soft clay soil is generated under long-term subway traffic load (Chen et al. 2002; Wang et al. 2003). According to some relevant field monitoring data, large axial deformation and surface settlement occurred in some tunnel segments of Line 1, Shanghai Metro (Lin et al. 2000), which bring a certain impact on the normal operation of the subway trains and threaten the safety of subway moving. The generation of settlements starts when the pore pressure and soil effective stress change (Tang et al. 2003a, b, 2005). Therefore, studying the dynamic responses of soils surrounding the subway tunnel plays a significant role in subway design, construction, and operation.

With respect to the soil dynamic characteristics under the subway load, two aspects should be considered. One is the dynamic responses of saturated soft clay under subway traffic load. It includes dynamic frequency response, soil dynamic response amplitude, and also the development in lateral and vertical directions. On the other side, it is the deformation response which is directly related to the settlement problem surrounding the subway tunnel. And this aspect is reflected by the continuous monitoring of pore pressure. The pore pressure response in saturated soft clay is much sensitive to the subway traffic load. The properties are various as to different soil types. Even for the same soil, the pore pressure varies distinctly as well due to different load properties. Previously, the pore pressure development characteristics of saturated sand are studied really in deep in all kinds of loads. Wang et al. (2004) analyzed the pore pressure properties in sandy mixture soil under cyclic load and established the relationship between pore pressure and energy loss. Zhang and Meng (2003) focused on the excess pore pressure in saturated sand due to impact loads. The critical impact strength to reach full liquefaction of sands is discussed. Zhang et al. (2003) simulated the earthquake response in clayey layer underlying silts and sands. The excess pore pressure increase was figured out by his simulation. Bai (2003) and Yang et al. (2002) computed and analyzed the pore pressure development in saturated layers under dynamic compaction load and blasting load, respectively. Okada and Nemat-Nasser (1994) demonstrated the reasonability of the pore pressure prediction by energy loss from the soil structure view.

The research on saturated soft clay is on the very starting stage as to sands due to the complex particle composition, physical properties, and pore pressure kinetic laws in soft clay. The accumulation and dissipation of excess pore pressure under traffic loading are both relatively slower. Xu et al. (1997) investigated the pore pressure increase of saturated soft clay under undrained cyclic loading through laboratory tests. Because of the cyclic traffic load, the pore pressure in clayey soils changed regularly with some special laws. And to figure out all these characteristics, continuous monitoring of pore pressure along the depth and also in different locations is needed for the relevant data analysis.

Previous scholars were concentrated in the theoretical analysis based on the laboratory tests. The field tests are really rare. Our research group declares that the most convincible analysis and laboratory tests should be on the basis of the field monitoring data, which can be the best descriptions and suitable to leading the mechanism study. Hence, in this chapter, pore pressure and earth pressure transducers were buried along the depth in boreholes and distributed laterally in different locations as well. As the subway advanced, the earth pressure and pore pressure were real-time monitored and recorded. The development of earth pressure and pore pressure under subway traffic loading was figured out based on the field data. The comparisons at various depths and along the lateral distances are both discussed.

2.2 Engineering Geology of Soft Soil in Shanghai

The soft soil is distributed widely, even omnipresent in Shanghai, which is typical soft sediment area located on the deltaic deposit of the Yangtze River estuary and formed in coastal marshes. The subsoil in Shanghai is composed of approximately 300 m loose Quaternary sediments with small regular variety at depths (Xu et al.

2.2 Engineering Geology of Soft Soil in Shanghai

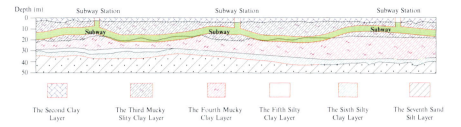

Fig. 2.1 The typical geological strata of Shanghai

2009), which consists of clay, loam, silt, and sand that vary from an estuarine to fluviatile sedimentation process. All the features and characteristics of these deposits show the physiographic evolution of the estuarine system. With the rapid development of economy and technology, particularly in the underground construction in Shanghai, the infrastructures were or are being constructed in the top shallow soft clayey deposit within subsurface 120 m, mainly consisting of Dark Green Stiff Clay (known as DGSC layer in subsurface around 70 m) with the following Upper Pleistocene (Q_3) and the Upper Holocene (Q_4) (Dassargues et al. 1991; Xu et al. 2009). Generally the soft soil in Shanghai is regarded as the layers upper the DGSC (Fig. 2.1) where the issues associated with human engineering activities are mainly concentrated. Moreover, engineering construction is much directly impacted by the third mucky silty clay layer and the fourth mucky clay layer. Meantime, the fourth mucky clay is the most occurrence layer embedded by the subway construction. In most areas, these layers are buried at a depth of 5 m. The thickness of mucky clay soil is relatively huge varying from 12 to 20 m with the absence of the third layer in some parts, which has characteristics of high water content, large pore ratio, low strength, and high compressibility attributable to the special information environment.

The model soil samples in this experimental research were obtained from Tang Town of Pudong New City in Shanghai, which is the typical soft clay soil construction site in Shanghai. In the studied zone, from a lithological point of view, the DGSC is characterized by a high bearing capacity, at a depth of about 70 m. It is overlain by a sandy layer usually called second aquifer. The latest unit formed in the estuarine environment is composed of sand deposits gradually replaced by fluvial silt and clay. The first aquifer is formed here and results in really shallow groundwater level mostly situated at a depth of 1–3 m. All the strata in the study area are shown in Fig. 2.1. The adverse engineering properties such as low permeability, thixotropy, and rheology of mucky clay soil are involved by many researchers over the years (Tang et al. 2003a, b, 2008, 2010; Xie and Sun 1996; Zhou et al. 2005), while the hydrothermal properties are rarely concerned.

2.3 Program Design

2.3.1 Selection of Testing Site

The field testing site was selected to the southern side of the east end in Jing'an Temple Station of Line 2 in Shanghai Metro. The subway train advances up to People Square and down to Zhongshan Park. The schematic is shown in Fig. 2.2.

2.3.2 Selection of Testing Instruments

The dynamic monitoring system was applied in the field test. The sampling frequency can reach as high as 200 Hz, and the precision is up to 0.1 kPa, which is very sensitive in the soil response to the subway traffic loads. The dynamic monitoring system consists of resistive sensors, dynamic strain amplifier, data acquisition, and computer. All the sampling data was recorded by the computer in real time. The whole process is depicted in Fig. 2.3.

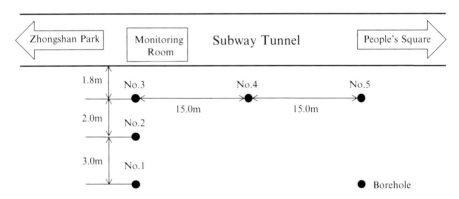

Fig. 2.2 Plan view of the testing site

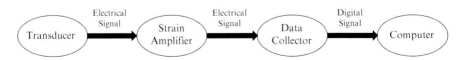

Fig. 2.3 Schematic diagram of dynamic monitoring system

2.4 Results and Analysis

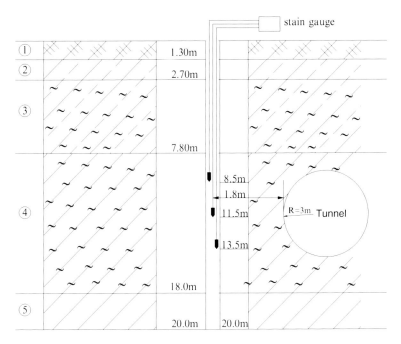

Fig. 2.4 Stratigraphic profile and instrument installation

2.3.3 Stratigraphic Section and Instrument Installation

As shown in Fig. 2.2, all the testing boreholes are distributed in line with 110 mm diameters. From a plan view, there are respectively three both parallel and perpendicular to the subway tunnel. No. 3, 4, and 5 boreholes are drilled 15.0 m apart in the tunnel axial direction with only 1.8 m offset to the tunnel segment. Perpendicularly, No. 1, 2, and 3 boreholes are distributed to study the development of soil dynamic responses to the subway traffic loads along the vertical direction. It is important to note herein that the distances between them are not always the same (shown in Fig. 2.2), which increased from 2 to 3 m. Viewed from stratigraphic section, the subway tunnel was located in the fourth layer of gray mucky clay. And all the monitoring points along the borehole are respectively 8.5 m, 11.5 m, and 13.5 m at the depths. Pressure transducers were buried in places exactly. The details of stratigraphic section and instrument installation were depicted in Fig. 2.4.

2.4 Results and Analysis

This section is to process and analyze the raw monitoring data. Some methodology is explained here. The response frequency of soft clay around the tunnel to subway traffic loads can be obtained by observing the time-domain waveform.

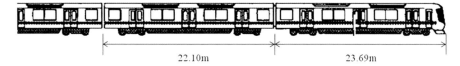

Fig. 2.5 The subway train (Model AC02) diagram of Shanghai Metro Line 2

Comprehensively observing each time-domain curve and the pore pressure development, the response characteristics of pore pressure at each depth can be figured out respectively. By comparison with the monitoring data at different measuring points along the depth, the wave propagation along the depth can be derived during the subway's operation. And the range of influence can also be simply predicted.

2.4.1 Soil Response Frequency

There are six cars in the subway train of Shanghai Metro Line 2 and the total vehicle length is 139.46 m (Fig. 2.5). The normal operation speed is 60 km/h with 30–45 s stop in each station. Since the testing site is much close to the station, the speed measured will be a little bit slower. According to the statistics, the speeds passing the measuring points are almost 30–40 km/h generally with 12–16 s duration. And in different periods, the intervals of the subway trains passing by the measuring points are different, generally 3–6 min and 3–4 min in rush hour times.

It can be easily seen in Fig. 2.5 that there are four concentrated (axle) loads in each subway car (two wheels in each end), which are transferred downward to the track structure and the foundation soil below through wheel rails. When the subway train approaches or passes the nearby measuring points, the traffic load spreads out by means of wave, and relevant response in soil can be observed.

In Fig. 2.6, the response wave when the subway train passed by one measuring point can be clearly seen. The response frequencies may be various due to different axle loads and track gauges. In a whole, two kinds of frequencies alternately showed up, which we called high frequency f_H and low frequency f_L. The most probable is that the high frequency caused by the loads of front wheels in one subway car and rear wheels in adjacent subway car periodically subjected to soils. The response time was relatively short. While as to the front wheels or rear wheels on each subway car, the cyclic loading time is longer and the response frequency is relatively low. Just as shown in the response wave figure, the soil vibration response can be easily found when each subway car approached and passed over.

To minimize surrounding impact on the data collection, 10 Hz filter was utilized in the field test. Numerous data were collected through continuous real-time monitoring. By comparing all the wave curves, a typical response wave of saturated soft clay was identified. Comprehensively summarized, conducted, and analyzed all the data, response frequencies at each depth of mucky clay soils were obtained

2.4 Results and Analysis

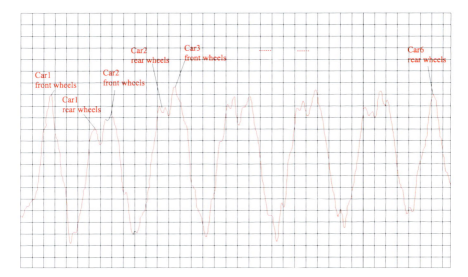

Fig. 2.6 The soil vibration response wave

Table 2.1 Frequency statistics

Measurement No.	High frequency (Hz)	Low frequency (Hz)	Measurement No.	High frequency (Hz)	Low frequency (Hz)
1-A-S	2.533	0.428	1-A-W	2.533	0.525
1-B-S	*	*	1-B-W	*	*
1-C-S	3.046	0.475	1-C-W	2.425	0.486
2-A-S	*	*	2-A-W	2.757	0.425
2-B-S	2.533	0.446	2-B-W	2.488	0.512
2-C-S	2.425	0.525	2-C-W	2.465	0.562
3-A-S	2.455	0.425	3-A-W	2.845	0.485
3-B-S	2.762	0.338	3-B-W	2.425	0.512
3-C-S	2.895	0.411	3-C-W	*	*
4-A-S	*	*	4-A-W	2.712	0.452
4-B-S	2.533	0.414	4-B-W	2.855	0.445
4-C-S	*	*	4-C-W	*	*
5-A-S	2.425	0.453	5-A-W	2.824	0.533
5-B-S	2.764	0.425	5-B-W	2.623	0.532
5-C-S	2.465	0.422	5-B-W	2.335	0.494

Note: In the Measurement No., the first number refers to borehole No.; *A*, *B*, and *C* refer to a depth of 8.5 m, 11.5 m, and 13.5 m, respectively; *S* refers to earth pressure transducer; *W* refers to pore water pressure transducer; * notes that no data were monitored in this borehole

(Table 2.1). Figure 2.7 also showed the histogram graph of response frequencies in each monitoring boreholes. Statistically, the high frequency f_H was in a range of 2.5–2.8 Hz and the low frequency f_L 0.4–0.6 Hz in value. Considering the most disadvantaged situation, 2.5 Hz was chosen in the laboratory tests.

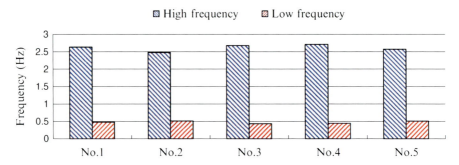

Fig. 2.7 Frequency histogram in monitoring boreholes

According to these two kinds of response frequencies from the field monitoring, combined with the natural frequency of saturated soft clay in Shanghai, some appropriate measurements should be considered to avoid similar or same frequencies between these two to resonance. If it happened, it would cause some engineering disasters such as the tunnel instability problems and segment cracking.

2.4.2 The Attenuation of Dynamic Response in Perpendicular Direction to Subway Tunnel Axis

Study on the subway traffic load involves many subjects, including locomotive dynamics, vehicle track dynamics, soil dynamics, and structure dynamics; subway tunnel is a partially enclosed space, in which subway train travels on the tracks. The wheel-rail contact force on the track structure is the main source of tunnel traffic load. It is always composed by two parts, the static load of rolling stock weight and vibration part by subway operation. The subway traffic load is generated by two aspects of both subway cars and track irregularity, in which the latter is the main reason.

In terms of vibration source mechanism, it has already a unified understanding, i.e., the high-speed operation of the subway train. It consists of two aspects of vehicles and tracks. As for the vehicle, the problem is mainly concentrated on wheels, such as wheel tread wear, wheel flat sliding, wheel out of round, and eccentric wheel, and about the track, when subway trains advance, a large number of wheel-rail contact forces conduct on the track simultaneously, which results in the vibration of wheels and track structure. Since jointless track and monolithic concrete bed are utilized in subway line, the track irregularity, especially the random irregularity, plays a great role on the vibration.

The study of attenuation development of soil dynamic response to subway load with the increase of distance from vibration source herein is firstly considered on the lateral direction. Previously, Chen et al. (1998) indicated experimentally that

2.4 Results and Analysis

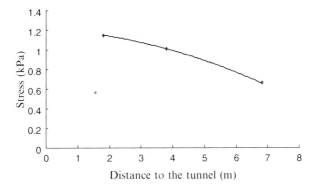

Fig. 2.8 Attenuation of lateral stress perpendicular to tunnel axis

Table 2.2 Fitting parameters

K_0	K_1	K_2
1.2122	0.0177	0.0093

the vibration was attenuated into zero at the distance of 25 m from the rail flange. Out of this range, soil dynamic response almost disappeared. Others also proved this point except the attenuation distance may vary with different traffic loads and circumstances. Meantime, buried deep in semi-infinite soils, the attenuation herein must be greater than surface. Seen in Fig. 2.1, No. 1, 2, and 3 boreholes were set in different distances from the tunnel sideline of 6.8 m, 3.8 m, and 1.8 m, respectively. To avoid the surface influences, all the transducers in different distances were located at the same depth of 13.5 m. Having comprehensively analyzed all the data in field, the results are shown in Fig. 2.8.

According to the field monitoring data and the analysis of the attenuation characteristics, the attenuation relationship with lateral distance perpendicular to tunnel axis is

$$\Delta p = K_0 - K_1 x - K_2 x^2 \tag{2.1}$$

where

Δp is the dynamic response value at distance of x (kPa)
K_0 is the dynamic response value at tunnel edge line (kPa)
K_1 is the first attenuation coefficient of dynamic response (kPa/m)
K_2 is the second attenuation coefficient of dynamic response (kPa/m^2)
x is the distance from the tunnel edge line (m)

Fitting all the monitoring data and combining regression analysis for Eq. (2.1), the parameters were obtained and presented in Table. 2.2.

Hence, the attenuation development with distances can be depicted in Eq. (2.2):

$$\Delta p = 1.2122 - 0.0177x - 0.0093x^2 \tag{2.2}$$

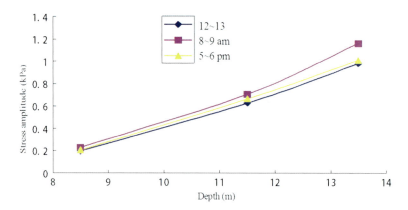

Fig. 2.9 Soil dynamic response amplitude with the depth

where, x is ranging from 0 to 11.33 m. With this formula the range of influence and the dynamic response values can be figured out. And the influence on surrounding environment can also be predicted and estimated on the basis of it.

From Fig. 2.8, as the distance increased, the soil dynamic response attenuation reflected on the dynamic stress amplitude. According to Eq. (2.2), the influence range can be calculated as 10.50 m. This result is distinct compared with the characteristics of the surface influence range of track traffic load. Besides the natural properties of soils, it's mainly because the shear modulus and damping ratio both increase along the depth.

2.4.3 The Dynamic Response Development Along the Depth

The transducers were distributed in soils around the top, axis, and bottom of subway tunnel. The early monitoring data indicated that the earth stress increased nearly linearly along the depth. During the subway operation, to figure out the relation of stress amplitude with the depth, rush hour was selected as the statistical duration for calculation. The statistical results were described in Fig. 2.9.

From Fig. 2.9, under the subway traffic loading, the biggest response was 0.23 kPa at a depth of 8.5 m, 0.7 kPa at 11.5 m, and 1.15 kPa at 13.5 m, respectively. The relation between the stress response amplitude and depth was almost linear. In addition, it can be also found that the dynamic stress responses were different in various subway operation periods, i.e., largest in the morning rush hour, second in the evening rush hour, and smallest at noon. It inferred that the passage flow during morning is largest and should be paid more attention in design and construction. The monitoring data during this period should be regarded as design reference.

2.4 Results and Analysis

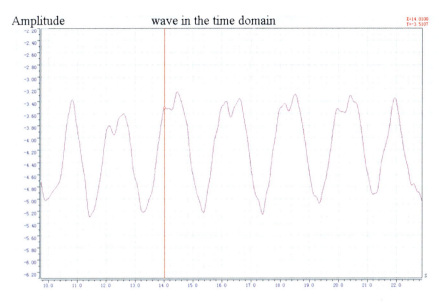

Fig. 2.10 Response curve of pore water pressure at subsurface 11.5 m

2.4.4 Development of Pore Pressure

Here taking the pore pressure development curve at a depth of 11.5 m as example to illustrate (Fig. 2.10), it described the pore pressure response curve during the subway operation, in which y-axis was the pore pressure response value (in $\mu\varepsilon$) and x-axis was time (in unit of s). According to the calibration coefficients, it can be transferred into curve of excess pore pressure verse vibration number (Fig. 2.10). With the same method, the pore pressure curves at the depths of 8.5 and 13.5 m in the location No.3 borehole were both conducted and the excess pore pressure curves were obtained (Figs. 2.12 and 2.13). From Figs. 2.11, 2.12, and 2.13, the excess pore pressure in surrounding saturated soft clay did not change a lot when the subway was advancing close, not exceeding 0.5 kPa.

The field monitoring data showed that the pore pressure dynamic response was caused by subway. It can be obviously seen that the pore pressure vibration number was just several and the amplitude was also very small. The subway operation interval was generally 3–5 min. During this time, the excess pore pressure can be almost dissipated completely. That is to say, the excess pore pressures induced by two adjacent subway cars would never superimpose. The details will be introduced in laboratory tests of the next chapter.

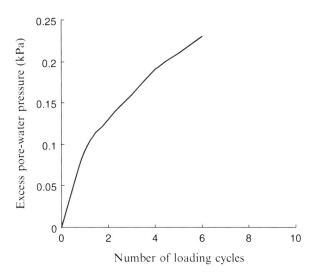

Fig. 2.11 Curves of over-static pore water pressure and vibration number at a depth of 11.5 m

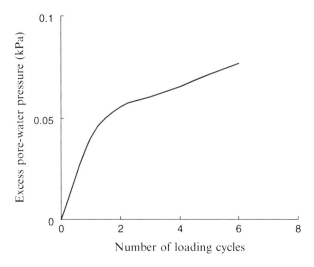

Fig. 2.12 Curves of over-static pore water pressure and vibration number at a depth of 8.5 m

2.4.5 Mechanism Analysis of Pore Pressure Development Under Subway Traffic Loading

The dynamic properties of pore pressure of saturated soft clay under subway traffic loading are characterized as follows: elastic compression occurred when the saturated soft clay subjected to traffic loads. It resulted in the rapid increase on pore pressure. After the subway passed by, the soil structure rebounded and negative

2.4 Results and Analysis

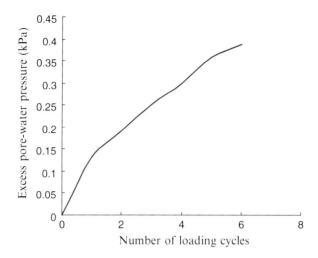

Fig. 2.13 Curves of over-static pore water pressure and vibration number at a depth of 13.5 m

pore pressure was generated and the total pore pressure decreased accordingly. As shown in Fig. 2.10, each subway train has six subway cars and each of them has four wheel sets with respective gauges of 2.8, 13.0, and 2.8 m. When each car wheel passed over the same measuring point, the increase and dissipation of pore pressure of saturated soft soil were both superposed. The dynamic response of pore pressure to the subway traffic load in each measuring point was related to the axle load transferred by wheels according to track structures. Therefore, the characteristics of pore pressure increasing and dissipating in each measuring point must correspond to the time and duration of the subway train wheel set passing each point. While times and intervals in different operation stages are various, and also there is some interference from another subway train, the pore pressure curve measured at each point is not the uniform wave curve. In the meantime, during rush hours, the amplitude of traffic load is bigger, corresponding energy transferred is greater, and the dynamic response of pore pressure is also relatively larger. All of this was reflected in Fig. 2.10. The subway traffic load was transmitted to the tunnel segment and lining structures and then transferred to surrounding soils by energy. After the interaction of soil and water, energy decayed gradually. Besides parts of energy absorbed by soils, some others were consumed by overcoming the initial hydraulic gradient of saturated soft clay to get the pore pressure increased. Once the traffic load disappeared, the pore pressure dissipated and water level recovered gradually. When the pore pressure did not dissipate completely before next train approached, the pore pressure fluctuated a lot constantly. When the subway stopped, the pore pressure dissipated and reached the final equilibrium steady state.

2.5 Chapter Summary

Subjected to the saturated soft clay surrounding tunnel areas of the Jing'an Temple Station-Jiangsu Road Station in Shanghai Line 2, continuous dynamic field monitoring was conducted by numbers of earth pressure and pore pressure transducers at different locations and depths. Some significant conclusions were obtained as follows:

1. During the subway train passing over, two dynamic response frequencies were generated in surrounding saturated soft clay. According to the measured dynamic response frequencies in soil and corresponding natural soil frequency, some appropriate measurements should be considered to avoid similar or same frequencies between these two to induce resonance, which will result in great damage to the subway tunnel.
2. In the direction perpendicular to the subway tunnel axis, the traffic load attenuated with increasing distance. The dynamic response attenuation with distance can be predicted in a formula by statistical analysis. The range of influence and the dynamic response amplitude were also obtained.
3. The dynamic response amplitude varied linearly with the depth. And the values in different rush hour stages were various, largest in the morning rush hours, secondary in the evening, and least at noon. It indicated that the passage flow was largest in the morning. During this stage, the soil was in a worst situation and the dynamic response values in this period should be chosen as the reference in the subway tunnel design and construction.
4. The pore pressure of saturated soft clay had very distinct response characteristics with the subway traffic load, and closer to the tunnel, the response is more sensitive.
5. The dynamic response of pore pressure in saturated soft clay had hysteresis effect. Besides the distance between measuring point and subway track, it also has some relation to the loading direction.
6. The pore pressure dissipation in saturated soft clay was much slower than the increasing time of pore pressure when subjected to the subway traffic load.

References

Alabi BA (1992) Parametric study on some aspects of ground-borne vibrations due to rail traffic. J Sound Vib 153(1):77–87
Bai B (2003) Pore water pressure of saturated stratum under dynamic impacts. Chin J Rock Mech Eng 22(9):1469–1473 (in Chinese)
Chen S, Xu GB, Gao R (1998) Dynamic analysis of the influence of train railway vibration on building along railways. J North Jiaotong Univ 22(4):57–60 (in Chinese)
Chen MY, Chen RP, Lu S (2002) Some soil mechanics problem in subway construction and operation in soft soil foundation. In: Proceedings of advanced technical forum on urban subway construction and environmental geotechnical. [Editor, Shi P D], Hangzhou, pp 165–177 (in Chinese)

References

Dassargues A, Biver P, Monjoie A (1991) Geotechnical properties of the quaternary sediments in Shanghai. Eng Geol 31:71–90

Hirokazu T, Shuhei S, Xie WP (2001) Train track-ground dynamics due to high speed moving source and ground vibration transmission. J Struct Mech Earthq Eng 682(7):299–309

Jones CJC, Sheng X, Petyt M (2000) Simulations of ground vibration from a moving harmonic load on a railway track. J Sound Vib 231(3):739–751

Lin YG, Liao SM, Liu GB (2000) A discussion of the factors effecting on longitudinal deformation of subway tunnel. Undergr Space 20(4):264–267 (in Chinese)

Lipen AB, Chigarev AV (1998) The displacements in an elastic half-space when a load moves along a beam lying on its surface. J Appl Maths Mech 62(5):791–796

Okada N, Nemat-Nasser S (1994) Energy dissipation in inelastic flow of saturated cohesionless granular media. Geotechnique 44(1):1–19

Pan CS, Li DW, Xie ZG (1995) Vibration effect of Beijing subway traffic on the environment. J Vib Shock 14(4):29–34 (in Chinese)

Sheng X, Jones CJC, Peryt M (1999) Ground vibration generated by a harmonic load acting on a railway track. J Sound Vib 225(1):3–28

Tang YQ, Huang Y, Ye WM et al (2003a) Critical dynamic stress ratio and dynamic stain analysis of soils around the tunnel under subway train loading [J]. Chin J Rock Mech Eng 22(9):1566–1570 (in Chinese)

Tang YQ, Ye WM, Huang Y (2003b) Marsh gas in shallow soils and safety measures for tunnel construction. Eng Geol 67:373–378

Tang YQ, Zhang X, Zhou NQ (2005) Microscopic study of saturated soft clay's behavior under cyclic loading. J Tongji Univ Nat Sci 33(5):626–630 (in Chinese)

Tang YQ, Yang P, Zhao SK, Zhang X, Wang JX (2008) Characteristics of deformation of saturated soft clay under the load of Shanghai subway line No. 2. Environ Geol 54:1197

Tang YQ, Zhou J, Liu S, Yang P, Wang JX (2010) Test on cyclic creep behavior of mucky clay in Shanghai under step cyclic loading. Environ Earth Sci 63:321–327

Wang CJ, Ji MX, Chen YM (2003) The additional settlement of saturated soft clay foundation under subway loading. In: Proceedings of 9th soil mechanics and geotechnical engineering of Chinese civil engineering. Tsinghua University, Beijing, pp 1118–1122 (in Chinese)

Wang QX, Jiro K, Jiro T (2004) Study on excess pore water pressures of sands mixed with clays under cyclic loading. Chin J Geotech Eng 26(4):541–545 (in Chinese)

Xie N, Sun J (1996) Rheology behavior of Shanghai clays. J Tongji Univ 24(3):233–237

Xie WP, Hu JW, Xu J (2002) Dynamic response of track-ground systems under high speed moving load. Chin J Rock Mech Eng 21(7):1075–1078 (in Chinese)

Xu CJ, Du J, Zhou HB (1997) Establishment of model of pore-water pressure of saturated soft soil under cyclic loading. Shang Geol 3:16–21 (in Chinese)

Xu YS, Shen SL, Du YJ (2009) Geological and hydrogeological environment in Shanghai with geohazards to construction and maintenance of infrastructures. Eng Geol 109:241–254

Yang B, Zhang ZY, Yang NH et al (2002) The analysis of the change of pore-water pressure in soft soil during blasting. Blasting 19(3):8–9 (in Chinese)

Zhang JF, Meng XY (2003) Build-up and dissipation of excess pore water pressure in saturated sand under impact loading. Chin J Rock Mech Eng 22(9):1463–1468 (in Chinese)

Zhang ZY, Lv XL, Chen YQ et al (2003) Experimental study on excess pore water pressure of soils covered by clay layer. Chin J Rock Mech Eng 22(1):131–136 (in Chinese)

Zhou XM, Yuan LY, Cai JQ, Hou XJ (2005) Analysis of soft soil distributed characteristics and deformation examples of Shanghai area. Shanghai Geol 4:6–10 (in Chinese)

Chapter 3
Laboratory Tests

3.1 Introduction

With rapid highway, subway, and light rail construction, there are numbers of incidents in consequence of large soil deformation induced by such long-time cyclic loading as subway vibration. Once after one trail operation of a Shanghai subway line, great influence on strength and deformation occurred in the soils surrounding the subway tunnel. It is largely relevant to the subway motion state. The subway vibration loading, which is different from static loading, and also varies from earthquake loading, is a kind of special cyclic loading characterized by reciprocating action in long-time periods. Hence, it has great engineering practice significance to study the dynamic properties and deformation controlling of soils surrounding the subway tunnel under vehicle loading.

At present, several scholars at home and abroad have been conducted researches on dynamic characteristics of soft clay soil under cyclic loading, and there are already some valuable research achievements. The study was mainly concentrated on pore pressure variation, accumulation deformation properties, strain softening characteristic, and dynamic elastic modulus attenuation and damage. Several factors such as loading rate, dynamic stress ratio, over-consolidation ratio, consolidation ratio, frequency, and static-dynamic stress deviator have been considered during the course of analysis.

Based on field investigation, a series of laboratory tests including CKC cyclic tri-axial tests, GDS dynamic cyclic triaxial tests, scanning electron microscopy (SEM) tests, and automotive mercury intrusion porosimetry (MIP) have been conducted to discuss in depth on pore water pressure, dynamic strength, dynamic stress-strain relations, and dynamic elastic modulus under various vibrated frequencies and vibrated numbers, in which the micro-deformation and damage mechanism were involved as well.

3.2 Pore Pressure Development

Pore water pressure properties under dynamic loading play significant role on variation of soil deformation and strength. It is the key for dynamic effective stress analysis method. Hereby, precisely predicting variations of pore water pressure is really important in our research. Presently researches are mostly focused on pore water pressure increasing and dissipation properties of sandy soils. Zeng et al. (2005) and Gong et al. (2004) studied the pore water pressure of slits and sands under cyclic loading, and pore water variation properties were presented. Zhou and Wang (2002) analyzed pore water pressure corresponding to stress and strain path and summarized pore water pressure variations during sand liquefaction under various factors. Shao et al. (2006) discussed the testing parameters which impacted the sand deformation behavior under cyclic loading. Li et al. (2005) analyzed mechanisms of silt liquefaction, impact factors, and pore water pressure variation during vibration. Guo et al. (2005) described undrained cyclic pore water pressure properties under various primary stress directions during vibration.

However the study on pore water pressure properties of saturated soft clay under subway loading is still on exploration stage. Soft clay has properties of large void ratio, high water content, poor permeability, high compressibility, and poor shear strength, which results in special and unique engineering behaviors. Since the particle composition, mechanical properties, and water transportation laws of soft clay are all different from sands, excess pore water pressures under vibrated loading increased or dissipated both relatively slower than the sandy soils. Under the subway loading, the pore water pressure development is greatly related to stability of subway line subgrade and surrounding building as well. As for the pore water pressure variation in saturated soft clay, Matasovic and Vucetic (1992, 1995) have already involved in this area at numbers of aspects. Clayey soils have strong regional properties, and the engineering behaviors would be distinct a lot because of various consolidation stats, loading modes. All these resulted in numerous pore water pressure models and there is no unified mode presented. Wu et al. (1998) considered high dynamic stress condition (cyclic stress ratio ≥ 0.05) and presented the pore water pressure properties under this situation. Zeng et al. (2001) depicted the pore water pressure increasing and dissipating properties of saturated soft clay under blast loadings. Meng et al. (2004) concluded several important results in dynamic pore water pressure response of saturated soft clay on blast loadings. Subjected to the subway loading, the pore water pressure properties of saturated soft clay can be hardly found.

Carrying out the study on dynamic deformation and damage mechanism of soft clay in depth provides significant theoretical and practical meanings on effective controlling axis deformation and mitigating regional land subsidence around the subway tunnel and also references for subway design, construction, and relevant subsidence prediction and prevention. The mechanical properties response, stress-strain characteristics, microstructure damage properties, and process are all involved herein.

3.2 Pore Pressure Development

Fig. 3.1 GDS dynamic triaxial test system

In this chapter, subjected to Shanghai mucky clay soil under subway loading, combined with the field monitoring data, GDS tests have been conducted according to the soft clay properties in Shanghai. The pore pressure increment model has been established via regression analysis, and the attenuation of growth rate of pore water pressure was studied as well. Finally the dissipation characteristics of pore water pressure were discussed by the energy loss principle.

3.2.1 Experimental Summary

3.2.1.1 Experimental Design

Apparatus

The dynamic cyclic triaxial test system, Geotechnical Digital Systems (GDS), which was produced by GDS Instruments Ltd in Britain, was utilized during the whole test (shown in Fig. 3.1). Automatically monitoring the entire testing process and recording all the real-time data at high speed, the GDS system is a set of fully digital experimental system characterized by high-resolution, easy operation, and reliable results. Figure 3.2 presents the schematic diagram of GDS.

Composed of triaxial pressure chamber, cell pressure controller, back pressure controller, axial loading driver, measuring apparatus, data collector, and computer (Fig. 3.2), GDS can meet the needs of consolidation test, standard and advanced

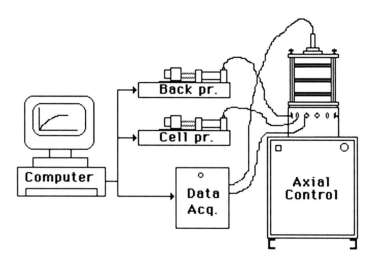

Fig. 3.2 Operation schematic diagram of GDS

triaxial test, stress path test, permeability test, and dynamic triaxial test, respectively, using different test software modules. Overall, it can be divided into three main parts, which are pressuring system, measuring system, and controlling system. Three pressure chambers provide triaxial pressure, confining pressure, and back pressure, respectively, during the tests. Back pressure chamber is used to control pore water pressure in testing samples and also can provide back pressure for the sample saturation. Cell pressure controller can apply confining pressure for samples. Axial load is subjected to sample by axial pressure chamber, which can also control the axial deformation (strain controlled and stress controlled are both achieved through this). All data of stress and strain, etc. are transmitted to the computer through sensors. All the functions in GDS are detailed as follows:

1. The radial and axial strain can be monitored and recorded in real time by Hall-effect sensors.
2. The pore water pressure can be read and recorded in real time with fast and high-precision characteristics.
3. The isotropic consolidation can be carried out automatically.
4. Cyclic triaxial tests can be conducted under frequency range of 0–5Hz, differing from various test sample moduli.
5. There are U-U, C-U, C-D, K_0, and stress-path testing modules.
6. All the data during the whole test can be real time stored, rendered, and even processed automatically, as shown in Fig. 3.3.
7. The graphical user interface makes the entire experiment under the full control.

3.2 Pore Pressure Development

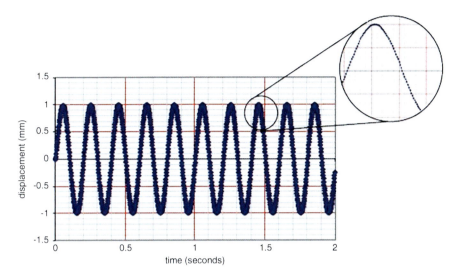

Fig. 3.3 Example of data analysis module of GDS

Experimental Design

The undisturbed saturated soft clay is under the natural stress condition, which is consolidated at K_0 state. When subjected to subway vibration loads, the amplitudes in these two directions could not be equal, and the frequency and amplitude of dynamic stress is rarely referred in previous researches. Hence, to investigate the soil properties under this irregular load, our monograph designed the laboratory experiments for further study, based on all parameter results from field monitoring tests.

Soil dynamic instrument is various all over the world, in which cyclic triaxial system is popularly used in China. In this experimental program, the advanced dynamic triaxial testing system GDS (Global Digital System) is utilized. Firstly, the cylindrical sample was back pressure saturated; to best simulate the field condition, K_0 consolidation was conducted, in which confining pressure σ_h was applied according to natural soil layers by the coefficient of lateral pressure at rest $\sigma_h = K_0 \sigma_v$; when the consolidation stage was finished, cyclic stress σ_d can be exerted accordingly. In this program, the dynamic loads were deduced from fielding data. Through the huge continuous monitoring data, combined the layers information in all boreholes, the response frequencies were concentrated in two sections: a high frequency of 2.4–2.6 Hz and a low frequency of 0.4–0.6 Hz. 2.5 Hz was chosen as the adverse circumstance to perform in this designation. In addition, under the subway vibration load, the response stress amplitudes along depth were processed. At a depth of 8.5 m, the maximum stress variation was 0.23 kPa, correspondingly 0.70 kPa in 11.5 m and 1.15 kPa in 11.5 m. The response stress amplitude behaved linearly as depth. And it was found that the amplitude values

Table 3.1 Scheme of undrained dynamic triaxial test

Test no.	Sampling depth (m)	Simulation depth (m)	Axial pressure (kPa)	Dynamic stress (kPa)	Frequency (Hz)
830D1	8.0–9.0	8.5	155	0.23	2.5
826D2	11.0–12.0	11.5	220	0.7	2.5
828D3	13.0–14.0	13.5	250	1.15	2.5

Table 3.2 Basic physical mechanic index of the 4th layer of saturated soft silt clay

Water content (%)	Void ratio (e)	Liquid index I_L	Plastic index I_p	Specific unit G_s	Coefficient of earth pressure at rest K_0	Coefficient of permeability (m/s)
51.3	1.44	1.28	21.6	2.75	0.58	3.00×10^{-7}

differed in various subway operation periods: morning rush hour duration > evening rush hour duration > afternoon rush hour duration. Considering the most adverse condition, the laboratory experimental design should take the maximum response amplitude into account.

The field information presented that, when subway advanced nearby, the dynamic response in the surrounding soil was periodically monitored. Hence, the stress-controlled cyclic loading module (dynamic triaxial test) was conducted. The specific controlling parameters were presented in Table 3.1.

3.2.1.2 Experimental Process

Test Preparation

Test samples used in the study were mucky clay drilled from the fourth soil stratum in Shanghai with a depth of 13.5 m near Jing'an Temple Station of Line 2 of Shanghai Metro. With a thickness of 10 m, the mucky clay stratum with mica and shell debris and thin silt has high natural moisture content, large void ratio, high compressibility, low strength, and medium-high sensitivity. Some physical index properties were presented in Table 3.2. The sampling depth was kept consistent with the simulation depth as shown in Table 3.1 to ensure the precise stress condition. All the samples were penetrated by the thin-wall sampler in size of 300Φ100 mm. From Fig. 3.4, it can be easily seen that the test preparation can be divided into several steps: specimen cutting, filter attaching, latex filming, specimen fixing (weighting), water injection for pressure chamber, and finally starting the tests.

1. Specimen cutting: the undisturbed 140Φ70 mm sample was cut into a specific cylindrical shape in size of 80Φ38 mm by a special soil sharpener (Fig. 3.4a). To avoid the influence of specimen variation on testing results, the sample should meet some requirements: (a) uniformity of soils, (b) water content is roughly the same, and (c) disturbance as little as possible.

3.2 Pore Pressure Development

Fig. 3.4 Preparation workflow before starting test. (**a**) Specimen cutting. (**b**) Filter attaching. (**c**) Latex filming. (**d**) Specimen fixing. (**e**) Water injection for pressure chamber. (**f**) Starting the test

2. Filter attaching: this step is to ensure the hydraulic connection on the specimen surface. It facilitates to penetration and drainage. As shown in Fig. 3.4b, there were three line filter paper attached around the specimen's body surface. Two pieces of circular filter paper were pasted at both end surface of the specimen with two small holes in each center for measuring the wave velocity.

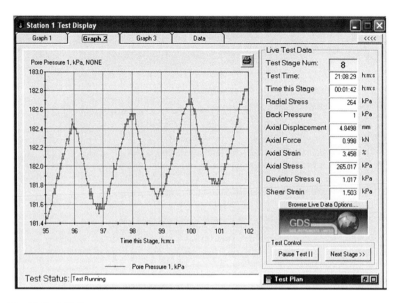

Fig. 3.5 GUI of GDS system

3. Latex filming: making latex membrane surround the specimen without much disturbance (Fig. 3.4c).
4. Specimen fixing: before fixing the prepared specimen, the weight should be measured for comparison after testing. The specimen was fixed in the axial force support pedestal (Fig. 3.4d).
5. Water injection for pressure chamber: for the accuracy requirement, purified water was used during the whole testing (Fig. 3.4e).
6. Starting the test: open the valves of back pressure and confining pressure. As shown in Fig. 3.4f, input the parameters needed and start the test. The GUI (graphical user interface) is presented in Fig. 3.5.

Test Procedures

1. Saturation stage: Back pressure was applied on the specimen utilizing the Satcon software module for more than 120 min to make the sample fully saturated. The pore water pressure coefficient B-value was measured to confirm the saturation until it reaches up to 0.98.
2. K_0 consolidation: To simulate the field stress condition, the sample was consolidated utilizing advanced triaxial test software module under the condition that the value of lateral pressure coefficient K_0 was about 0.7, which lasted for more than 24 h till axial strain rate was less than 0.05 %/h.
3. Dynamic triaxial test: After the completion of consolidation, the shear wave velocity was measured before starting the dynamic triaxial tests. Relevant

parameters such as frequency, amplitude, the specimen's stiffness, number of cycles, and testing termination conditions. Then cyclic triaxial test was conducted continuously on the sample with advanced triaxial test software module under step loading, in which the initial cyclic stress amplitude was 10 kPa, the periodical amplitude increment was 5 kPa, and the number of loading cycles in each step stage was 2,000 times. All tests were conducted according to the program designation in Table 3.1.
4. When the test was finished, the specimen was taken out and weighted and finally covered with plastic wrap to store in a container for the microstructure testing.

3.2.2 The Pore Water in Soft Clay

3.2.2.1 Electric Double Layer

In clayey soil, the clay content is relatively high; and clay particles have characteristics of colloids or sols. The water in soil pore structure is not the pure water. So the soil-water system actually is a soil-water-electrolyte system. The theoretical basis to study the interaction of soil-water-electrolyte system is the electric double layer theory in colloidal chemistry.

Resulting from any combination of the following factors, isomorphous substitution, surface dissociation of hydroxyl ions (hydrolysis), and absence of cations in the crystal lattice and absorption of anions and presence of organic matter, the surfaces of clay particles carry a net negative charge. This net negative charge of the mineral attracts cations in the water to the surface of the mineral. And the attracted cations become permanently bonded to the surface of a clay particle; it is said to be "fixed" and termed as potential-predominant ions. Additional water molecules are also attracted toward the clay minerals. The cations are hydrated, in which the water molecules, being polar, are clustered around the cations in a nonrandom manner to form a hydrated water film.

After the interaction of mineral with charge and water solution, the cations in hydrated water film attracted to a clay mineral surface also try to move away from the surface because of their thermal energy. A diffusion tendency is generated by the net effect of the force due to attraction and repulsion, i.e., the forces of attraction decrease exponentially with an increase in distance from the clay particle surface, as shown in Fig. 3.6. The first layer of this double layer system is formed by the negative charge on the surface of the soil colloid (clay) particle. The cations are almost rigidly attached close to the clay particle surface and move together with clay particle during ionization. This layer is termed as rigid layer. The outer layer of second part away from the clay particle with ions of the opposite charge (cations) is called diffuse layer.

The theory of ionic distribution surrounding the colloid particles was firstly proposed by Helmholtz. He conceived the electric double layer in colloid was

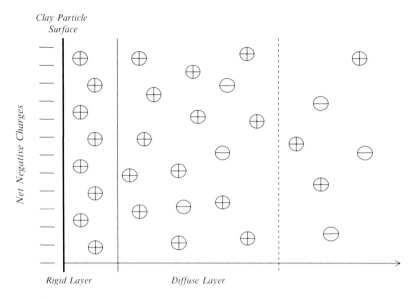

Fig. 3.6 Electric double layer

similar to the interaction in a parallel plate capacitor. The distance between these two layers is a molecule diameter. The potential between the dielectrics decreased linearly.

Gouy (1910) termed the outer part of the double layer "diffusion" because the ion concentration decreased with the increase in distance from the surface of the clay particle, according to an exponential law. The Gouy-Chapman diffuse double layer did not consider the influence of molecule diameter.

Stern model is a variation of the above model that considers cations have a finite size and that some ions may be selectively adsorbed onto the clay surface as a closest layer, and the negative charge of clay particles is balanced by the sum of the positive charges in the Stern and diffuse layers. He showed that neither the Helmholtz nor Gouy-Chapman theory alone is adequate. He presented a model combining the essential features of both of them, including a quasi-Helmholtz strong interior layer and a Gouy-Chapman outer diffuse layer. Thickness of interior layer depends on the size of attracted cations. The potential within this layer declines linearly. And the potential in outer diffuse layer exponentially decreases with the distance to the particle surface, as shown in Fig. 3.7. The solid particle is moving together with surrounding layer, rather than separately. The total potential of this double layer is the potential difference between the solid particle and the normal liquid interface, termed as thermal potential (ε). The rigid layer has opposite charge to the diffuse layer. The potential difference between these two layers can only arise when the relative movement between particle and hydrated solution occurs, called electrokinetic potential (ζ).

3.2 Pore Pressure Development

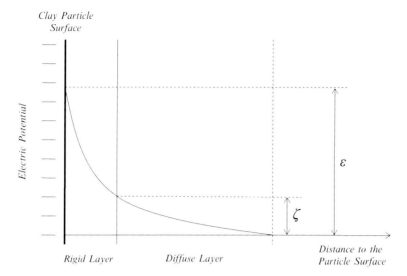

Fig. 3.7 The variation of potential in relation to distance

3.2.2.2 Pore Water in Saturated Soils

The water in the electric double layer consists of hydration water (ionized water) and molecule water. Depending on the bonding ability of clay particle surface, the water can also be divided into strongly bound water and loosely bound water. The strongly bound water is termed as adsorbed water (hygroscopic water) as well. It refers to the water held by electrochemical existing on the soil surface. The strong attraction between these surfaces causes an extremely thin water film (a few molecules in thickness), in which these water molecules completely lost the mobility. Under the influence of electrical forces, its properties are different from the normal water. The mechanical properties are more akin to an almost solidified state. It is much more viscous and elastic, and its surface tension and shear strength are also greater. But it could not pass the water pressure and has no electrical conductivity and ability of dissolving. As mentioned above, as for the thickness of adsorbed water film, most scholars regard it is several molecules thick. Some researchers also opine that it must be different in various locations of mineral particle. The highest thickness can be up to hundreds of molecules.

The loosely bound water is also known as thin film of water, which is slightly away from particles in the outer layer of the strongly bound water. It is composed of the main part of the water film. The orientation and adsorption of the loosely bound water is much weaker than that of the strongly bound water. And the thickness varies a lot. Some scholars think it is just 20–30 molecules thick, while it was measured as hundreds even thousands of water molecules sometimes. The variation of thickness impacts the physical and mechanical properties of fine soils. It depends on the

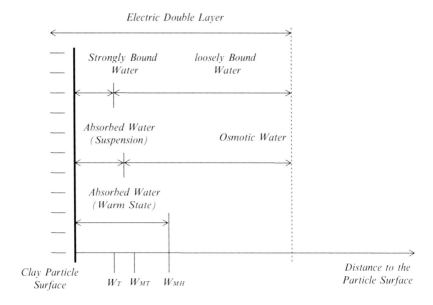

Fig. 3.8 Schematic diagram of bound water on the clay particle surface

particle size, shape, and mineral composition, and also PH value of the electrolyte solution, the cations, and concentrations.

The electric double layer merges into the normal water in a relatively far distance from the clay surface, where the water molecules are hardly affected by the electrostatic attraction. They are mainly controlled by gravity and keep in free mobility. Corresponding to the bound water, the water in this part is called non-bound water or free water.

The surface bound water in clayey soil is presented in Fig. 3.8. It facilitates to show the connection of water content and the soil structure. In the range of completely dry to adsorbed water content W_T, the interior adsorbed water is formed strongly attracted surrounding the clay particle surfaces. During the initial stage of the interaction between clay particle and water solution (to W_T), the strongly bound water is deployed in Volmer-Weber mode. When the water content reaches the maximum adsorbed water content W_{MT}, the amount of strongly bound water grows sharply while forming the capillary water. As the adsorption of water continues, the loosely bound water is firstly formed in the pores. Then, the strongly bound water and loosely bound water both augment. There is a transitional stage; the formation of adsorbed water ends up when the water content approaches the maximum water molecule concentration W_{MT}. From all aspects, the loosely bound water occurs at the beginning of W_{MT} and is most actively formed at the plastic limit water content W_P. Several different understandings indicate that the growth of loosely bound water ends up at liquid limit water content W_L or maximum expansion water content W_{MH}. Since the loosely bound water results in the soil volume expansion, it is regarded

3.2 Pore Pressure Development

that the loosely bound water approaches the highest thickness when close to W_{MH}. W_{MH} is the largest water retention. It is the distinction of strongly bound water and loosely bound water, which is also the maximum critical distance controlled by the surface clay particle negative charge.

3.2.3 The Pore Water Development Properties

Soils usually consist of particles, water, and air, as a multiphase porosity media, in which particles make up solid body skeleton and air and water are substances in gas and liquid phase filling in the pore structure surrounding all soil particles. As for saturated soils, pores are full with water without air. According to conventional consolidation theory, the external loads are carried by pore water when instantaneously applied to the soil. With the consolidation advancing, the external loads are gradually transferred to the effective stress $\sigma - \mu$, which the soil strength and deformation mainly depend on. Hence, for the mechanical properties of soils, the pore water pressure development under external loads should be figured out first of all. As we know, the undrained pore water pressure is induced by the variation of soil volume. The earliest study was empirically related to the variation of pore water pressure with external loads, i.e., the so-called stress theory, such as Skempton and Henkel equations. Skempton equation (1954) had clear concept and simple formula, and the parameters can be determined by laboratory tests. It was widely used in geotechnical calculations. But the parameter A or B is not a simple constant, relating to the stress-strain properties of soils. Lo (1969) pointed out the pore water pressure could not be represented only by stress. Through a series of laboratory tests, an exclusive relation between pore water pressure and major principle strain existed, regardless of stress-strain path, loading time, number of loading cycles, and drainage conditions. In fact, pore water pressure was a strain effect. Henkel, and Cai et al. performed further relevant research and respectively proposed the pore water pressure equations of saturated and unsaturated soils under stress invariant expression. And a pore water pressure function taking account of the soil's stress-strain relation was established by triaxial tests. Therefore, the pore water pressure is not only relevant to external loads and also greatly influenced by the soil's stress-strain properties. Study on variation of the pore water pressure can help to understand the distribution of pore water pressure, effective stress, and their transferring circumstance in soils under external loads. It can make significant sense to further acknowledge the deformation mechanism in the soil.

In Fig. 3.9a–c, the pore water pressure linearly ramped fast after the onset of external loads. And then the curve bended as the increasing velocity slowed down. While the absolute growth value was still relatively large, after a period of growth, the pore water pressure approached a stable stage and got into the maximum value. In these three group tests, they are 115, 160, and 185 kPa, respectively.

Combining the above description and Fig. 3.9, it can be easily found that the pore water pressure development under cyclic loads can be divided into three stages,

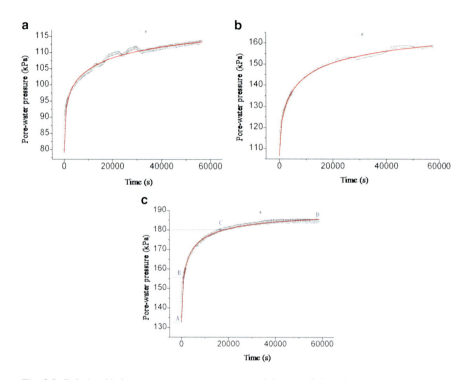

Fig. 3.9 Relationship between pore water pressure and time: (**a**) 8.5 m; (**b**) 11.5 m; (**c**) 13.5 m

including fast ramping stage, slow growth stage, and stable stage. The detail was specifically depicted in Fig. 3.9c as follows:

Fast Ramping Stage (AB)
When applying the cyclic loads, the excess pore water pressure was generated and ramped in a line and approached 50 % of maximum value quickly (approximately 1,300 s, 3,200 cycles), while the increase of pore water pressure decayed very soon.

Slow Growth Stage (BC)
This stage was a long transition period, in which the excess pore water pressure passed over an abrupt increased state into stabilization. During this stage, the pore water pressure value was still increasing, but the rate relatively slowed down.

Stabilization Stage (CD)
During this stage, the pore water pressure hardly increased even with the cyclic load continuously applied. It stabilized at a constant value, which was the maximum limit under this stress level.

Combining previous engineering projects' information and the soft clay properties, the experimental data were fitted and logistic-type pore water pressure model (Eq. 3.1) was proposed (all correlation coefficient reached above 0.99).

3.2 Pore Pressure Development

Table 3.3 Model parameters

Test no.	u_0	u_t	N_0	p	Correlation coefficient
D1	78.96156	128.21944	8,125.42544	0.43157	0.9933
D2	106.44487	172.99848	6,383.8695	0.590636	0.9978
D3	132.94314	191.73222	2,130.01121	0.64803	0.9978

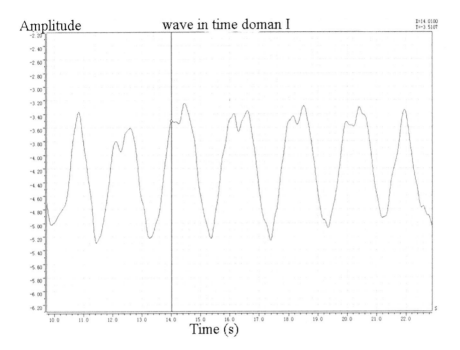

Fig. 3.10 Curve of in situ pore water pressure at a depth of 11.5 m

$$u = u_t - \frac{u_t - u_0}{1 + (N/N_0)^p} \qquad (3.1)$$

where N was the number of cycles, u_0 was average static pore water pressure value, u_t was excess pore water pressure limit value, and N_0, p were the regression parameters, obtained from the regression analysis of test data. All results were shown in Table 3.3.

According to the field monitoring information, the above model was fitted as well. Figure 3.10 presented the variation of pore water pressure investigated in a depth of 11.5 m when a subway train advanced nearby, in which the y-axis was the pore water pressure response value (kPa) and x-axis was time (s). All the original monitoring data were transferred into the curve of pore water pressure verse number of vibration cycles according to the calibration coefficients of transducers (Fig. 3.11).

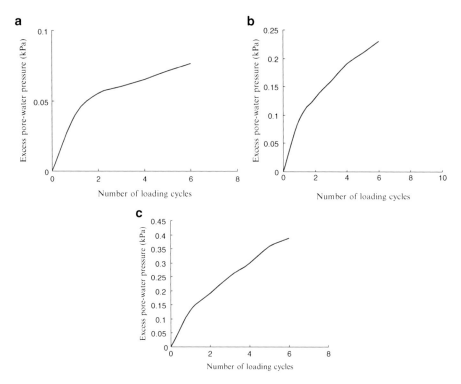

Fig. 3.11 Curve of over-static pore water pressure and vibration number. (**a**) 8.5 m. (**b**) 11.5 m. (**c**) 13.5 m

Comparing Fig. 3.9 with Fig. 3.11, it can be seen that the growth rate of pore water pressure in field monitoring data was slower than tested in laboratory and the increment was smaller as well. The main reason was opined to be the size effects and the condition difference of field testing. Hence, parameter C was introduced into Eq. (3.1) to modify the pore water development model. The new model is shown in Eq. (3.2).

$$u = u_f - \frac{u_f - u_0}{1 + C(N/N_0)^p} \tag{3.2}$$

where C was the modification parameter.

The field monitoring data in depths of 8.5, 11.5, and 13.5 m were fitted to determine the value of C, and results were presented in Table 3.4.

Modified curves and fielding data are both presented in Fig. 3.12. From the curves, it can be seen that the modified pore water pressure value by Model (3.2) was much close to the field monitoring value. It indicated that the pore water model can be used to predict the pore water pressure during the subway's moving around.

3.2 Pore Pressure Development

Table 3.4 Values of fixed parameters C

Test no.	C
D1	0.0350
D2	0.2112
D3	0.2975

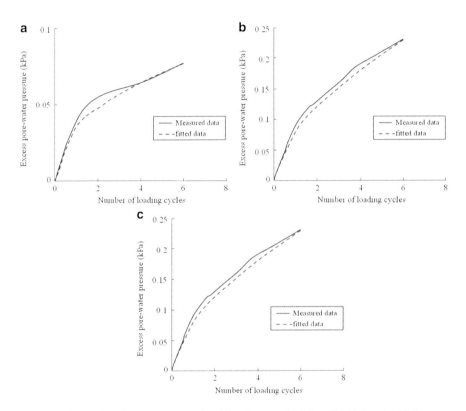

Fig. 3.12 Comparison between measured and fitted curves. (**a**) 8.5 m. (**b**) 11.5 m. (**c**) 13.5 m

And then using effective stress theory, the deformation of soft clay surrounding the subway tunnel can be dynamically analyzed. This can provide guidance for the subway operation.

The field measurement data indicated that the excess pore water pressure generated by the vibration load was not large in amount. And the subsequent dissipation was really fast as well. The duration of subway moving interval was 3–5 min. During this time, the pore water pressure can be almost fully dissipated, i.e., there was no conspicuous superposition effect induced by adjacent subway train operation. The vibration load in this circumstance is equivalent to just several cycles in dynamic triaxial test. During this duration such that the pore water pressure developed only as the first stage in Model (3.2), fast ramping stage.

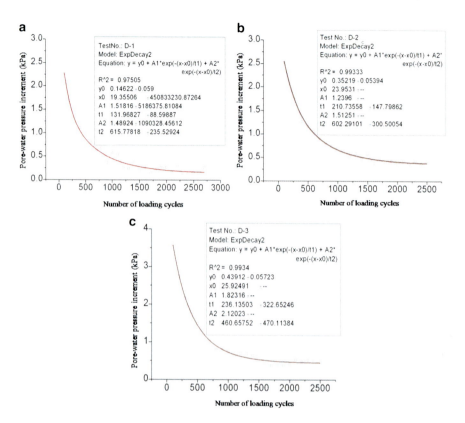

Fig. 3.13 Curves of increase rate of pore water pressure

3.2.4 The Pore Water Pressure Attenuation Properties

Figure 3.13a–c presented the ramping velocity curve of pore water pressure in the first stage (AB). Each data point was the pore water pressure increment after a certain number of vibration cycles. From the figures, during the fast ramping stage, the pore water pressure was not increasing at a stable rate as a line. The ramping rate was slowed down sharply and then got into decaying. After the regression analysis, the ramping rate of the pore water pressure can be fitted by ExpDecay2-type curves as Eq. (3.3)

$$\Delta u = \Delta u_0 + A_1 e^{-(N-N_0)/\zeta} + A_2 e^{-(N-N_0)/\eta} \qquad (3.3)$$

where Δu was pore water pressure ramping rate, N was the number of vibration cycles, and $\Delta u_0, A_1, A_2, N_0, \zeta, \eta$ were all regression parameters. The specific values were detailed in Table 3.5.

3.2 Pore Pressure Development

Table 3.5 Attenuation model parameters of pore water pressure

	Δu_0	A_1	A_2	N_0	ζ	η	R^2
830D1	0.14622	1.51816	1.48924	19.35506	131.96827	615.77818	0.97505
826D2	0.35219	1.2396	1.51251	23.9531	210.73558	602.29101	0.99333
828D3	0.43912	1.82316	2.12023	25.92491	236.13503	460.65752	0.9934

3.2.5 The Mechanism Analysis of Influence of Subway Vibration Loads on Pore Water Pressure

There are six cars in each subway train of Shanghai Line 2 and the total vehicle length is 139.46 m. The normal operation speed is 60 km/h with 30–45 s stop in each station. And the subway vibration loads were applied lasting 20 s when it was moving nearby the monitoring points. In the different operation periods, the intervals of subway passing by varied around 3–6 min. In the morning and evening rush hour periods, the intervals were both 3 min and the train axle loads were relatively larger and the energy propagated from wheel axles were bigger. It stands to reason that the pore water pressure response variation amplitude was relatively larger. Hereby from the view of this point, the most adverse circumstance was considered in this laboratory experimental program.

During the subway operation, the energy generated by subway vibration was passed down to the surrounding soils by the subway track. The energy gradually decayed during the propagation in soils. Accordingly, disturbed by dynamic energy, the water and soil particles were both deformed correspondingly. Compared to the soil particles, the water in the pore structure of soils was much more sensitive. And at the beginning of vibration, the most energy was absorbed by the pore water and resulted in the abrupt increase in pore water pressure. As the effective stress applied was reciprocated up and down on the soil particles, the chemical bonding among these particles was loosened and broken down as energy transferred from water to soil particles. Until the main energy dissipated among soil particles, the pore water pressure gradually approached into a dynamic equilibrium.

With the disappearance of vibrations, the excess pore water pressure started to dissipate, and the water level recovered down again. From the field measurement data, the excess pore water pressure generated by the passing-by subway was not really large. And the induced vibration was just several cycles loading. The subsequent dissipation was really fast as well. The interval of subway operation was 3–5 min. During this time, the pore water pressure can be almost fully dissipated, i.e., there was no conspicuous superposition effect induced by adjacent subway train operation. After the operation of the subway stopped, the reciprocated movement of pore water pressure ceased and the equilibrium finally approached.

3.3 Deformation Characteristics of Saturated Soft Soils

Even after long-term consolidation process, differential settlements will occur in the soft foundation when subjected to continuous traffic loads. With rapid development of urban rail transit, great economic and social benefits are boosted. The environmental impacts could not be ignored, such as in Lines 1 and 2 of Shanghai Metro and Line 1 of Nanjing Subway; from the design depth, it can be found that most subways were located through the saturated soft clay layers. The settlement in a certain section in Shanghai Metro exceeded 20 cm, which severely affected the subway operation, and some crack arose in the surrounding old buildings. The influence of subway vibration on the subway line and surrounding environment has paid more attentions currently. Since the soft clay has characteristics of large void ratio, natural water content, poor permeability, high compressibility, and low shear strength, it behaves some exclusive engineering properties. The factors influencing deformation and strength are various and complex. Hence, further study on the dynamic properties, deformation, and damage mechanism of soft clay is still needed, not only to effectively control the axis deformation of subways, avoid or alleviate the surrounding land subsidence disaster, and provide references for the subway design, construction, and the prediction for the subway settlement but also to acquire the physical and mechanical responses of soft clay to different subway vibration loads, stress-strain properties, and microstructure damage characteristics and evolution process. This research program plays a significant role in the theoretical value and practical meaning.

The deformation characteristics of soft clay are really complicated under cyclic vibration loads. Generally, the relation of stress-strain is expressed from three aspects of nonlinearity, hysteresis, and accumulation effect. A lot of scholars conducted many researches on the dynamic properties of soft clay under cyclic vibration loads. While the previous work was mainly concentrated in analysis on the influences of dynamic stress and vibration loading cycles on the residual deformation, pore water pressure, and strain softening, the study on accumulated strain characteristics of soft clay is still far from enough. Monismith et al. (1975) proposed an exponential model, in which the accumulated plastic strain was relevant to vibration loading cycle number.

$$\varepsilon_p = A_0 N^b \tag{3.4}$$

where ε_p was accumulated plastic stain, N was the cycle number of vibration loading, and A_0, b were the relevant testing parameters. Equation 3.4 mainly explained the influence of the loading cycle numbers, and the other factors were considered in parameters A_0, b. Parr (1972) figured out a residual strain model by taking into account the accumulated plastic strain and loading cycle number in the cyclic triaxial tests.

3.3 Deformation Characteristics of Saturated Soft Soils

$$\lg\left(\frac{\dot{\varepsilon}_N}{\dot{\varepsilon}_1}\right) = \lg C_0 + \varsigma \lg N \tag{3.5}$$

where $\dot{\varepsilon}_N$ was the plastic strain rate after the nth cyclic loading, $\dot{\varepsilon}_1$ was plastic strain rate after the 1st cyclic loading, and C_0, N were both the experimental parameters. When N equals to 1, C_0 was 1. Jiang (2000) conducted the experimental study on settlement, strain rate of soft clay under cyclic loads. He considered the influences of loading rate, dynamic stress ratio, and over-consolidation ratio on the residual strain rate. But his work did not mention any of static deviator stress and dynamic deviator stress. Huang et al. (2006) introduced relative deviator stress parameter to modify Eq. (3.5) based on critical soil mechanical theory, taking initial static stress, cyclic stress, and undrained limit strength into account. The undrained accumulated strain in soft clay was figured out under combination of static and cyclic loading.

This section mainly discussed the accumulated strain characteristics under cyclic loads, by considering the number of loading cycles, frequency, amplitude, and confining pressure. Herein CKC cyclic triaxial apparatus, GDS, and the accessory shear wave velocity testing instrument (GDS BES) were all used.

3.3.1 Experimental Introduction

3.3.1.1 Experimental Instruments

All the dynamic triaxial tests in this part were performed on the CKC cyclic triaxial testing system, which is a unidirectional-excitation cyclic loading apparatus produced in the USA. In this instrument, cyclic tests can be conducted by applying a stress-controlled harmonic excitation loading at a loading frequency of 0–2 Hz on a cylindrical sample in size of 80Φ39.1 mm, and the confining pressure can be controlled in a range of 0–1.2 MPa. The maximum cyclic strain was within 10^{-5}–10^{-2}; the maximum lateral exciting load was 2,500 kN. There are three kinds of wave form that can be applied in this apparatus: sine, square, and saw-tooth wave forms. The instrumentation of this apparatus equipped three testing systems of dynamic stress, dynamic strain, and pore pressure, including five sensors: external load cell, external LVDT, and three differential pressure transducers. Dynamic elastic module, dynamic damping ratio, and liquefaction strength can be measured.

The loading mechanism in this cyclic triaxial system can be introduced firstly (Chan 1981). A dual channel electropneumatic system is used to convert the electronic signal to pneumatic pressure, which in turn applies the axial load and lateral pressure. An electronic-to-pneumatic (E/P) pressure transducer coupled with a volume booster relays to pressurize the actuator. A function generator provides an electronic signal of the desired load trace to a moving coil in the E/P transducer, which is a force-balance instrument. A force is produced on the input coil by the interaction of the input current and the permanent magnetic field. An increase in the current to the coil creates a downward thrust, which in turn restricts the flow air out of a nozzle, causing an increase in the output pressure. The force of this output pressure acting over the area of the nozzle balances the force generated on the input coil. Within a range up to 100 kPa, the output pressure is directly proportional to the change in DC current. A pneumatic amplifier is used to amplify the relatively low pressure and very limited low capacity by 1:4 ratio of diaphragm areas. Moreover the amplified output pressure in air was further increased by a volume booster relay by a factor of two. Finally the output pressure has a large flow capacity and applies on the actuator.

The testing principle of cyclic triaxial tests is introduced as follows: during the cyclic triaxial test, firstly, the sample is consolidated under a desired confining pressure σ'_0, which is determined according to natural stress condition of soil layers. The applied cyclic load should best simulate the potential maximum acceleration that may be subjected on the soil, self-weight of all layers, and additional loads in buildings; hereby the equivalent cyclic stress σ_d can be calculated. The cyclic load is applied on the sample by means of semi-peak amplitude. Hence, during each loading cycle, an amplitude of $\sigma_d/2$ is applied; the sample is generated a normal stress of $\sigma'_0 \pm \sigma_d/2$ in the 45-degree slope surface, in which the cyclic shear stress is $\tau_d = \pm \sigma_d/2$. The cyclic triaxial test utilized the shear stress in 45-degree slope surface to simulate the shear stress of τ_d on the horizontal surface in the soil at the subway tunnel bottom in field under the subway vibration load.

3.3.1.2 Experimental Sample

The sample used in this experimental program was undisturbed soils. They were obtained near the Sanshan street station of Nanjing subway, which was grey muddy silt clay at the depth of 10 m underground. It is saturated with low plastic, medium compressibility. There are thin silt layer embedded, containing iron-manganese nodules. The thickness was about 4 m. The specific physical and mechanical indexes were presented in Table 3.6.

3.3.1.3 Experimental Designation

The dynamic loads subjected to the soil are mainly relevant to the excitation load amplitude, frequency, subway speed, and acceleration. Hence, in this experiment, the loading amplitude σ_d, loading frequency f, and confining pressure σ_3 were considered during the simulation. Generally the vibration frequencies of subway are in a range of 5–15 Hz. Many research results indicated that the dynamic properties of soils were rarely influenced by loading frequency at low range of 0–20 Hz. Limited by the instrument, frequencies of 0.5, 1, 1.5, and 2Hz were designed. And the loading amplitude was characterized in terms of the cyclic stress ratio (CSR = $\sigma_d/2\sigma_c$), which for the cyclic triaxial tests is the ratio of the maximum peak shear stress to the initial consolidation confining stress (σ_3 or σ_c). Herein the CSR values were employed as 0.1, 0.2, 0.3, and 0.4. In addition, different consolidation

3.3 Deformation Characteristics of Saturated Soft Soils

Table 3.6 Basic physical mechanic index of test soil

Soil	Water content ω (%)	Natural density ρ_0 (g/cm^3)	Dry density ρ_d (g/cm^3)	Saturation Sr (%)
Mucky Silty Clay	50	1.75	1.23	95

Void ratio	Direct shear parameters		Shear compressibility	
e	C (kpa)	φ°	a_{1-2} (Mpa)	m_v (Mpa^{-1})
1.21	12	15	0.475	4.5

degrees and different confining pressures were considered as well. According to the field geology survey, the static coefficient of lateral pressure in silty sand was 0.45 and 0.46 in mucky clay.

To explore the permanent deformation, dynamic elastic module, and dynamic strength properties of saturated mucky clay in Nanjing Subway, a series of 20 cyclic triaxial group tests, two resonant-column tests, and two direct shear tests were conducted.

3.3.1.4 Experimental Procedures

Testing Preparation

To precisely reflect the deformation behaviors of the soil under cyclic vibration loads, the field condition should be simulated best correctly. The simulation for reality refers to soil properties, initial stress conditions, subway vibration loading, and drainage conditions.

1. Soil properties: This is realized by using undisturbed soil samples. In this experimental program, the soil sample preparation was really perfect. Obtained from the construction site by thin sampler, the undisturbed soil was sealed in the sampling tube by plastic tapes. Several layers of thick sponge pad were placed surrounding the sampling tubes to prevent disturbance on the way from Nanjing to Shanghai. After the specific desired sample were made out, they were set into a container for saturation by using vacuum pump. Generally vacuum pumping for 2 h and water saturating for around 10 h, an undisturbed sample was achieved.
2. Initial stress conditions: When field site was horizontal, isotropic consolidation was performed to simulate the zero-original shear stress condition. If the field site was slope surface, anisotropic consolidation was used to simulate original shear stress condition τ_0. The coefficient of the lateral pressure at rest K_c was considered in three levels of 1.0, 0.44, and 0.46. The consolidation pressure was the natural self-weight stress or marginally larger.
3. Subway vibration loads: The subway vibration load was characterized in terms of vibration amplitude and frequency. The stress condition of the soil in the subway tunnel bottom was influenced by the overlying structure and subway moving speed. According to relevant researches, when the subway advanced at speed

of about 150 km/h, the dynamic stress almost hardly changed. When the speed exceeded 150 km/h, the dynamic stress increased proportionally to the speed. When the speed reached 300 km/h, it approached a stable value again.

In addition, from the field monitoring data, the frequency of subway loads differed continuously. It mainly depended on the overlying structure, subway train type, and advancing speed. To simply evaluate the subway vibration frequency, disturbance distance L was introduced into the calculation as $f = v/L$, where v was the subway moving speed and L was a parameter related to soil depth. When calculating the soil near the ground surface, L can be taken as the largest distance between the wheel sets of two adjacent subway cars, i.e., 7.0–8.0 m. According to this method, the vibration frequency can be 5–15 Hz in the soil just beneath the subway tunnel. The track here referred to steel rail. The situation was much similar to the circumstance of forwarding shield. The vibration generated by subway train was a unidirectional pulse wave on track or near the track. After the decaying on the track, it was approximate to a constant-amplitude sinusoidal wave. Hence, in this program, a harmonic sinusoidal vibration applied to simulate the dynamic loading can make clear sense.
4. Drainage conditions: At the moment of subway vibration instantaneously applied to the saturated soil, almost no water dissipation or volume change happened. Undrained conditions were employed in these triaxial tests.

Testing Procedures

1. Sample preparation: A standard 80 cm height, 39.1 cm diameter cylindrical specimen was shaped in specific soil sharpener and then placed in a vacuum chamber for water saturation by air pumping. The soil sample should be pre-saturated for 24 h in water and then air pumped for 2 h. In addition, when removing the vacuum pump, the specimen was still needed to saturate in the water for 10 h before used in the tests.
2. Fixing the specimen: After full saturation, the specimen was latex filmed by a specific mold. Firstly the latex membrane was preset inside the sample mold and tightly attached around the walls by a rubber pipette bulb. Then, the specimen was placed into the mold and fixed in the cyclic triaxial system.
3. Applying confining pressure: The value depended on the natural self-weight stress or made to be marginally larger. If a deviator stress was needed, it can be applied in stages to reach full consolidation (when the pore water pressure dissipated totally).
4. After complete consolidation, the drainage valves should be closed. The vertical dynamic load σ_d was applied under the undrained condition, keeping with the same confining stress. This cyclic load was the subway vibration load. Then, the real-time cyclic stress σ_d, cyclic strain ε_d, and dynamic pore water pressure u_d were depicted in curves and can be observed by GUI system in the computer.

3.3 Deformation Characteristics of Saturated Soft Soils

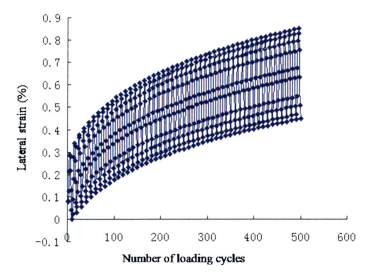

Fig. 3.14 Deformation of soft clay

3.3.2 *The Factors on Deformation Under Cyclic Loads*

The deformation of soft clay under cyclic loads consists of elastic deformation and plastic deformation. The plastic deformation was also called residual deformation. It refers to deformation value remaining in the soil after removing the dynamic loads. In the dynamic triaxial tests, it was characterized by the ratio of specimen's height difference after applying dynamic loading to the original height. The residual deformation is the unrecoverable deformation after dismounting the dynamic loads. As shown in Fig. 3.14, the elastic deformation was recoverable, while the plastic deformation was unrecoverable and has accumulation effect. With the increase in number of vibration loading cycles, the elastic deformation became larger and the accumulated plastic deformation increased as well, but the growth rate declined. All factors affecting the residual deformation were discussed as below.

3.3.2.1 Cyclic Stress Ratio (CSR)

To discuss the influence of CSR on soil deformation properties under cyclic loads, four CSR levels of 0.1, 0.2, 0.3, and 0.4 were considered in this experimental program. The variations of dynamic strain ε under different CSR values were plotted in Fig. 3.15.

The above curves were the accumulated strain of muddy silt clay increased with the vibration cycle number under a loading frequency of 2 Hz, in different CSR values. The results indicated that the dynamic strain ε accumulated with vibration time

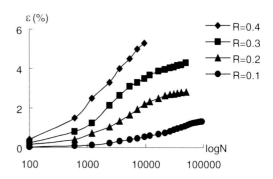

Fig. 3.15 Curves of dynamic strain and vibration number under different CSR

at the same consolidation condition and confining pressure. A critical CSR value existed in muddy silt clay, which was around 0.3–0.4. This critical value divided the dynamic strain behaviors into two types: attenuation mode and failure mode.

3.3.2.2 Frequency

The residual strain variations under different loading frequencies were presented in Fig. 3.16. It can be easily seen that in the range of low frequency, the influence on deformation behaviors was not conspicuous. From $\varepsilon - N$ curves, during the same vibration time, the deformation induced by low-frequency loading was relatively larger, compared to high frequency. The results revealed that the muddy silt clay behaved more elastic properties under high-frequency loading. Since when the loading frequency was high, the pore water pressure generated in the soil could not be dissipated in time, and longer loading time may result in more fully pore water pressure dissipation. The creep behavior must be more obvious. Hence, the soil behaved more plasticity and viscosity effects.

In addition, when CSR was smaller than 0.3, the deformation curve was in the attenuation mode. It was mainly characterized as recoverable elastic deformation, and the plastic deformation rarely accumulated. At the initial loading stage, the deformation ramped quickly. And then it gradually compacted and the strain increment declined until reaching a stable value. That is to say, in the initial operation period, the dynamic strain increased with the vibration time. But this increment would gradually decrease at the later period until it approached a stable value, when the subway vibration loads have no significant influence on the soil deformation. Under this circumstance, ε and N presented a semi-log relation, as shown in Eq. (3.6).

$$\varepsilon = \alpha + \beta \log N \tag{3.6}$$

where α, β were parameters related to dynamic stress level and soil properties and N was number of loading cycles. Through regression analysis, it can be obtained that when CSR = 0.1, $\varepsilon = 0.23 \ln N - 1.45$; when CSR = 0.2, $\varepsilon = 0.52 \ln N - 2.69$; the coefficient of correlation was high as 0.9539.

3.3 Deformation Characteristics of Saturated Soft Soils

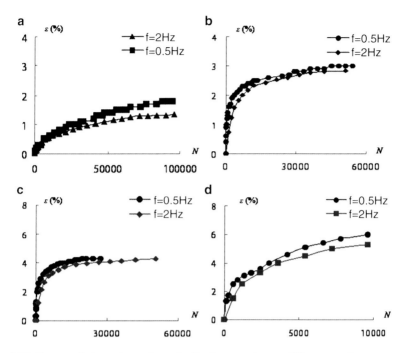

Fig. 3.16 Curves of dynamic strain and vibration number at different cyclic stress ratio. (**a**) CSR = 0.1. (**b**) CSR = 0.2. (**c**) CSR = 0.3. (**d**) CSR = 0.4

When CSR was larger than 0.4, the sample behaved as failure-type deformation curve. Dynamic strain value augmented with the vibration time. After a certain periods, the structure of the sample was damaged by the reciprocated load. And the failure rate became faster as an increase in CSR. In this situation, ε and N presented an exponential relation as Eq. (3.7):

$$\varepsilon = aN^b \tag{3.7}$$

where a, b were parameters related to dynamic stress level and soil properties and N was number of loading cycles.

When CSR = 0.4, $\varepsilon = 0.04N^{0.56}$; the coefficient of correlation was 0.9677.

A lot of scholars all over the world attained the similar conclusions, such as Matsui (1980), who conducted dynamic triaxial tests on clay with $I_p = 55$, under loading frequencies of 0.02–0.5 Hz. Results showed that pore water pressure and lateral deformation increased as vibration time lasted. For a certain vibration duration, larger pore water pressure and lateral deformation were generated under low loading frequency. Zhang and Tao (1994) performed cyclic triaxial tests on saturated soft clay with $I_p = 21.6$ and concluded that lowering the loading frequency, higher pore-water pressure and lateral strain can be resulted. Zhang et al. (2006) utilized the saturated anti-seepage clay from an earth dam and did the dynamic

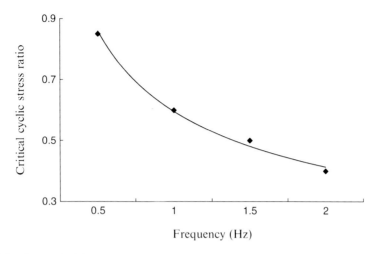

Fig. 3.17 Curves of critical cyclic stress ratio verse vibration number

triaxial tests. He also pointed out the deformation developed farther when applied low frequency. And the dynamic elastic modulus and damping ratio increased with frequency.

Vibration frequency reflects the subway speed in some extend. Under the same consolidation condition and the same confining pressure, there existed a critical cyclic stress ratio. To explore the adverse condition that the soil was easily damaged, the relation between frequency and critical CSR was figure out in Fig. 3.17.

From the regression curve in Fig. 3.17, the frequency had great influence on critical cyclic stress ratio. When the loading frequency was low, the critical cyclic stress ratio was relatively large; and the value decreased with higher frequency. Until the frequency increased into a certain value, the critical CSR was rarely changed any more. It can be explained that when loads were applied, the structure of the soil sample was damaged and reflected in volume change and resulted in deformation. After removing the loading, some elastic deformation was recovered and took some duration. Lower frequency permitted the soil structure to recover at a relatively longer duration. When the frequency approaches a higher value, the soil structure could not recover in a short time with the load continuously applied. The critical CSR approached stabilization.

3.3.2.3 Consolidation State

The dynamic properties change accordingly when disturbed by subway operation. In this experimental program, different consolidation time was performed to simulate the various influence degree when the soil was subjected to cyclic vibration loads. Under deviator stress consolidation conditions, 2 and 5 h were

3.3 Deformation Characteristics of Saturated Soft Soils

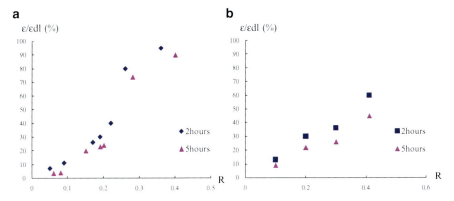

Fig. 3.18 Soil deformations at different consolidation state. (**a**) $K_c = 1$ and $N = 1,000$. (**b**) $K_c = 0.46$ and $N = 1,000$

conducted, respectively. The deformations after a certain vibration loading cycles were compared under the same CSR condition to investigate the strength variation with time development, as shown in Fig. 3.18.

From Fig. 3.18, consolidation time influenced the deformation properties under the same stress condition. The deformation in the sample with 2 h consolidation was much greater than generated in 5 h. And as the CSR was enlarged, this affection was conspicuous and the nonlinearity behaved severely. It implied that longer the consolidation time, more stabilized the soil structure became and smaller deformation will be resulted.

3.3.2.4 Consolidation Confining Pressure

The study on the residual deformation of soils in the various depths is rarely seen. In this experiment, the different depths were characterized in terms of consolidation confining pressure, since the geological conditions were really complex and the coefficient of the lateral pressure differed. In addition, according to the subway geological survey information and the overall plan in first phase of Nanjing Subway construction, the depth varied even in a same subway line, such as near the shielding construction field of Diaoyutai Station, the subway line extended from surface into underground, with large variation in the burial depth. Low burial depth referred to small confining pressure, and even in the same burial depth, the confining pressure differed if the coefficient of the lateral pressure changed. Thus, different consolidation ratios and confining pressure were necessary to discuss in this research work. In this part, consolidation ratios of 1.0 (isotropic consolidation), 0.6, and 0.46 were considered, corresponding to 130, 80, and 60 kPa in the confining pressure. All the triaxial cyclic tests were conducted under a loading frequency of 1 Hz with a loading amplitude of 40 kPa.

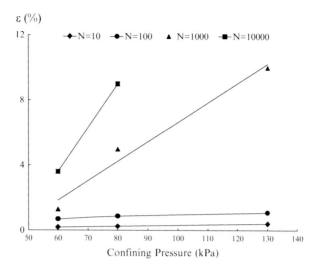

Fig. 3.19 Curves of dynamic strain and confined pressure in different vibration number

The dynamic strain variation with the confining stress was presented in Fig. 3.19. The growth rate augmented with the vibration time. This reflected that the relation of dynamic strain and confining pressure is influenced by dynamic stress level and loading time. When the dynamic stress conducted is the same, higher the confining stress, larger dynamic strain will be resulted and the sample is more easily to failure. The reason may be mainly because under the same consolidation lateral pressure, smaller confining pressure generated a larger initial shear stress at the beginning of loading. Hence, the deformation developed was not so easy in the further loading time. Accordingly, the relevant dynamic strain was relatively small.

3.3.3 Summary

According to all the factor analysis on the dynamic properties of muddy silt clay in Nanjing subway under vibration loads, some conclusions can be drawn as below. Firstly, a critical cyclic stress ratio has been found during this experimental program. The value is related to the soil properties, the dynamic loading, consolidation confining pressure, etc. When the dynamic load was smaller than the critical cyclic stress, the soil deformation increased as loading time developed. And the growth rate decreased correspondingly and finally approached a stable value. In contrast, when the dynamic load exceeded this critical cyclic stress, the soil deformation ramped quickly till it reached failure. Secondly, under the same other vibration condition, low loading frequency resulted in larger deformation, while relatively smaller in higher frequency loading. The critical cyclic stress ratio declined with increasing loading frequency and there existed an asymptotic line. Thirdly, under the same consolidation lateral pressure and the same dynamic load, after a certain vibration time, larger permanent deformation was generated in higher consolidation ratio, i.e., large confining pressure, in which the sample more easily got into failure as well. The reason may be a larger initial shear displacement occurred at the beginning of deviator stress consolidation. Hence, the reciprocated shear displacement was relatively smaller under the dynamic loads than generated in the soil through isotropic consolidation. As for muddy silt clay, longer consolidation time behaved more elastic properties, and accordingly the dynamic strain was relatively small.

3.4 Strength Characteristics of Saturated Soft Soils

The dynamic strength of soil refers to the dynamic stress when failure deformation occurred under reciprocated dynamic load. Large dynamic load σ_d with fewer loading times N or small dynamic load σ_d under more loading times can achieve the same dynamic strength in the tested soil. Therefore, the study of soil dynamic strength should be related to a specific loading time and soil deformation. The soil properties and loading mode (frequency) must be considered as well. And the failure deformation amount should depend on the specific requirement in projects.

All the dynamic triaxial tests were conducted in CKC cyclic triaxial system. The specific experimental conditions were detailed in Sect. 3.2.1. Herein the dynamic loading frequency was set to be 1.0 Hz and 5 % ε_p as the failure strain. Then, the natural logarithm relation of dynamic stress σ_d and loading times were presented in Fig. 3.20a. From the figure, the dynamic shear strengths can be derived at points of $N = 10, 100, 1,000$, and $10,000$. The specific value of each point is shown in Table 3.7. From the above data, the natural logarithm relation of dynamic shear strength τ_{df} and loading time is shown in Fig. 3.20b.

The static shear strength index of mucky clay was $C = 0.12$ kg/cm^3, $\varphi = 20°$. Combining the data in Table 3.7, it can be easily found that the dynamic strength was much larger than static strength under cyclic vibration load. As a matter of fact, the variation in internal friction angle was not so conspicuous. From Fig. 3.20b, the dynamic shear strength declined with increasing vibration loading time; and the falling rate gradually slowed down and finally the shear strength got into stabilization. This was much consistent with the previous analysis that the permanent deformation increased with loading time when the loading stress was smaller than critical dynamic stress. According to the above circumstance, the determination of strength parameters of soil beneath the subway tunnel in designation should be chosen depending on the subway vibration loads and loading times. Therefore, the engineering project designation should consider the specific requirement and determine the reasonable failure strain and choose the rational strength index. From this experiment, it can be known that when $K_c = 1, \varepsilon_p = 5$ %, and $\sigma_3 = 0.6, 1.0$, and 1.4 kg/cm^3, φ_d was measured to be in the range of $8°-12°$, and C_d was in the range of $0.03-0.1$ kg/cm^3.

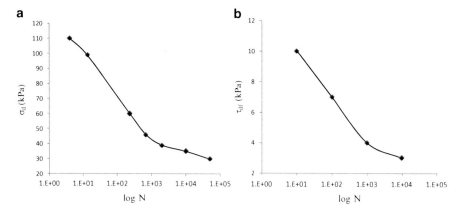

Fig. 3.20 The natural logarithm relation of dynamic responses and loading time: (**a**) dynamic stress; (**b**) dynamic shear strength

Table 3.7 Dynamic strength index at $K_c = 1$, $\varepsilon_p = 5\%$

Vibration time (N)	Lateral stress σ_1 (kPa)	Confining stress σ_3 (kPa)	Dynamic stress σ_d (kPa)	Dynamic cohesion C_d (kPa)	Dynamic internal friction φ_d (°)
10	118	60	58	10	12
	177	100	77		
	241	140	101		
100	110	60	50	7	10.8
	168	100	68		
	232	140	92		
1,000	101	60	41	4	11
	158	100	58		
	220	140	80		
10,000	95	60	35	3	8
	150	100	50		
	214	140	74		

Note: The coefficient of the lateral pressure was measured as 0.46 in laboratory

3.5 Creep Behavior of Soil Under Cyclic Loading (Tang et al. 2010)

Soil is a porous medium composed of solid particles, water, and air. Soil particles constitute the basic skeletons of soil. Water and gas fill the pores. Water and soil particles form strongly bound water and loosely bound water because of physical and chemical effects. Under external forces, the internal pores of the soil decrease and then accordingly deform. Water and air in the pores are being squeezed out at the same time. But the friction between the soil particles and pore water in saturated soft soil impeded the discharge of pore water so that the soil's deformation delays. Creep phenomenon of soil is induced by the soil viscous property, especially for

3.5 Creep Behavior of Soil Under Cyclic Loading (Tang et al. 2010)

saturated soft clay. Creep phenomena of clay impact the deformation and strength of soil. And for the engineering practice such as soil consolidation, slope stability, and ground bearing capacity, it cannot be ignored. Therefore, the creep characteristics of clay have been an important research content of soil mechanics.

In recent years, with the rapid development of urban transport, aircraft, subway, and light rail transport are widely used. While bringing convenience to the residents, it also triggered a series of engineering environmental geological problems. According to the field investigation in Saga Airport, the subsequent settlement caused by traffic loading has reached 15 cm (Miura et al. 1995). The tunnel settlement along axial line in some part of Shanghai subways has exceeded 20 cm (Wang and Liu 2000). The creep characteristics of clay under long-term cyclic loading are becoming a hot research topic. Many scholars have done researches on the creep behaviors and made some achievements. Since the early 1960s, there are many researches on the creep behaviors of clay under cyclic loading. Seed and Chan (1961) investigated the low embankment deformation under reciprocating traffic loading. In order to construct the gravity offshore platform, Andersen et al. (1980) conducted comprehensive research on Drammen clay characteristics under wave loading. Considering the element of pore water pressure dissipation, Hyodo et al. (1992) studied the pore water pressure and residual settlement of saturated soft clay in the subgrade under traffic loading. A model for predicting the cyclic behavior of soft clay was presented by Zhou and Tu (1996), while the one-dimensional sedimentation of clay under long-term cyclic loading was estimated by Jiang and Chen (2001) from China. Moreover, in the last decade, the frequent occurrence of subway accidents all over the world resulted that more and more attentions have been paid to various soil behaviors under the cyclic reciprocating loading. Chai and Miura (2002) presented a method for predicting the traffic-load-induce settlement of road on soft subsoil with a low embankment. Moses et al. (2003) carried out a series of cyclic triaxial shear tests to study the influence of the strain effect and load cycles on the undrained shear strength of cemented marine clay. Tang et al. (2003) conducted the dynamic triaxial test of silt soils around the tunnel in Nanjing under subway train loading and presented the variations of critical dynamic stress ratio and dynamic strain with time. Subsequently, He (2004) also studied the dynamic strength behavior and obtained several significant conclusions. Yılmaz et al. (2004) focused on the deformation behavior and undrained cyclic shear to consider the severity of foundation displacements. Boulanger and Idriss (2007) gave a kind of method to evaluation the potential for cyclic softening. Guan et al. (2009) estimated the rheological parameter for the prediction of long-term deformations in tunneling. However, most of these researches are focused on the residual deformation prediction in the long-term cyclic loading, and little attention was paid to the essential characteristic and variation of clayey creep. Hence, in order to avoid or control the engineering geological disaster caused by subway operation, it is indispensable to make experimental researches on the essential properties of clayey creep under cyclic loading.

We can see from the above that most of these researches focus on the residual deformation prediction in the long-term cyclic loading, but rarely on the essential characteristics of cyclic creep of soft clay. Hence, the purpose of our research is to provide more accurate and useful reference to predict deformation settlement of clay and establish rheological model under cyclic load, to avoid the engineering geological disaster. We performed a series of cyclic triaxial tests for undisturbed samples by step loading method. These tests designed on base of the field test results showed in the last chapter were to study cyclic creep of clay, and pore water pressure changes during the creep by considering load stress level and number of loading cycles.

3.5.1 Test Apparatus and Samples

The dynamic cyclic triaxial test system GDS (Geotechnical Digital Systems), which was produced by GDS Instruments Ltd in Britain, was utilized during the whole test. Composed of triaxial pressure chamber, confining pressure controller, back pressure controller, axial loading driver, measuring apparatus, data collector, and computer, GDS can meet the needs of Satcon and consolidation test, standard and advanced triaxial test, stress path test, permeability test, and dynamic triaxial test, respectively, using different test software modules.

Test samples used in the study were mucky clay drilled from the fourth soil stratum in Shanghai with a depth of 13.5 m near Jing'an Temple Station of Line 2 of Shanghai Metro. With a thickness of 10 m, the mucky clay stratum with mica and shell debris and thin silt has high natural moisture content, large void ratio, high compressibility, low strength, and medium-high sensitivity. Basic physical index properties were presented in Table 3.2. Manual of GDS and the preparation of samples in detail can be referred from Sect. 3.2.1.

3.5.2 Test Control Parameter and Procedures

Axial sinusoidal cyclic loading was applied on the samples under step loading during the tests to simulate the vibration loads generated by transport such as subway. We studied soil creep characteristics by the model of step loading. Before the formulation of test program, all test control parameters should be determined first.

The stress state of the undisturbed saturated mucky clay is in K_0 consolidation stress condition under natural stress conditions. In order to simulate the field conditions more accurately, the sample must be consolidated under K_0 condition before the application of vibration loads to restore to the field stress state. According to the physical and mechanical indexes of soil samples and engineering experiences, K_0 coefficient is 0.7 in the test.

The pressure set values in the consolidation process can be calculated. Back pressure, 100 kPa, promotes soil sample to reach saturation as quickly as possible

3.5 Creep Behavior of Soil Under Cyclic Loading (Tang et al. 2010)

Table 3.8 Control parameters of cyclic triaxial tests

Test plan	Vibration waveform	Loading frequency (kPa)	Axial pressure (kPa)	Confining pressure (kPa)	Back pressure (kPa)
C	Sinusoidal	0.5	235	195	100
D	Sinusoidal	2.5	235	195	100

Test plan	Basic pressure (kPa)	Initial stress amplitude (kPa)	Stepped stress amplitude (kPa)	Stepped loading vibration times
C	265	10	5	2,000
D	265	10	5	2,000

in the consolidation process. According to $\sigma'_1 = \gamma' h$ and $\sigma'_2 = \sigma'_3 = K_0 \sigma'_1$, we can know that effective axial pressure σ'_1 is 135 kPa and effective confining pressures σ'_2 and σ'_3 are both 95 kPa. So the values of axial pressure and confining pressure are 235 kPa and 195 kPa, respectively.

Considering the additional stress at the bottom of the tunnel by the train and tunnel system is between 20kPa and 40 kPa (depth of tunnel axis is 11–14 m), the basic value of cyclic loading is determined as 265 kPa. Changing the stress amplitude of cyclic loading, the value of vibration loads can be changed. The test uses the step loading mode; the initial stress amplitude of the sinusoidal cyclic loading is 10 kPa. Every level of cyclic stress amplitude has an equal increment of 5 kPa.

In addition, according to the conclusion of the subway field test in the last, the frequencies of the cyclic loading are set to be two representative frequencies 0.5 and 2.5 Hz in order to simulate metro vibration loading accurately.

Comprehensively considering the above controlled parameters, laboratory test program is presented in Table 3.8.

3.5.3 Test Procedures

Test procedures are as follows:

1. Installation of samples: The sampler was cut into a cylindrical specimen of 38 mm diameter and 80 mm height, which was fixed into the triaxial pressure chamber before the test.
2. Saturation under back pressure: Back pressure was applied on the sample utilizing the Satcon software module for more than 120 min to make the sample fully saturated. The pore water pressure coefficient B-value was measured to confirm the saturation until it reaches 0.98.
3. K_0 consolidation: To simulate the field stress condition, the sample was consolidated utilizing advanced triaxial test software module under the condition that the value of lateral pressure coefficient K_0 was about 0.7, which lasted for more than 24 h till axial strain rate was less than 0.05 %/h.

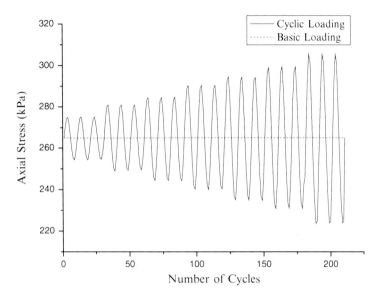

Fig. 3.21 Step cyclic loading during the cyclic triaxial test

4. Cyclic triaxial test: Cyclic triaxial test was conducted continuously on the sample with advanced triaxial test software module under step loading, in which the initial cyclic stress amplitude was 10 kPa, the periodical amplitude increment was 5 kPa, and the loading cycles in each step stage were 2,000 times. The whole cyclic loading process is shown in Fig. 3.21.

3.5.4 Composition of Clayey Creep Under Cyclic Loading

3.5.4.1 Creep Characteristic Curve

The characteristic analysis of mucky clayey cyclic creep curve under cyclic loading in Shanghai could be useful to study cyclic creep characteristics. Step loading method was used during the tests, and the samples have very similar creep behaviors during each step, whose curve trends are almost same. A typical creep curve would be shown as follows.

The control parameters of C sample during the first step loading are as follows: frequency 0.5 Hz, cyclic stress amplitude 10 kPa, and loading cycles 2,000. The cyclic loading during this stage is described in Fig. 3.22. Clayey creep of this stage is more obvious, and the cyclic loading stress amplitude is close to the train vibration load. So this stage can be defined as typical loading stage. Take a typical stage, for example, and do detailed analysis of clayey creep curve's characteristics during the first five cycles in the stage mentioned above.

3.5 Creep Behavior of Soil Under Cyclic Loading (Tang et al. 2010)

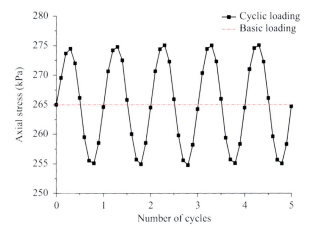

Fig. 3.22 Typical sinusoidal cyclic loading

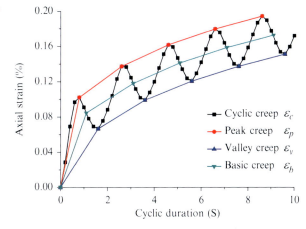

Fig. 3.23 Initial cyclic creep characteristic curve during the typical stage

Figure 3.23 shows the cyclic creep curve of C sample during the first five cycles in the stage mentioned above. In each cycle, ten data points were collected by the computer automatically. It can be observed that the creep curve gradually drifts up in an approximately sinusoidal waveform and the rate of movement gradually slows. There are a peak point (ε_p) and a valley point (ε_v) of strain in each loading cycle. After linking each peak point and valley point of strain in each loading cycle, respectively, we obtain two new curves, which are defined as peak creep ε_p curve and valley creep ε_v curve, which are shown in Fig. 3.23. Then, based on the test datum and the formula $\varepsilon_b = (\varepsilon_p + \varepsilon_v)/2$, basic value of strain is obtained. Then, connecting the point of basic value of strain in all the cycles, another creep characteristic curve can be obtained, that is, the basic creep ε_b curve. Therefore, there are three creep characteristic curves in the cyclic creep curves of saturated mucky clay in Shanghai. They are peak creep ε_p curve, basic creep ε_b curve, and valley creep ε_v curve. These three curves have the same variation trend which will be discussed in the following passages.

3.5.4.2 Analysis of Clayey Creep in Typical Loading Stage

Take typical loading stage, for example. Comparing the variations of the soil in the first five and last five loading cycles, we got some basic creep characteristics under cyclic loading.

The characteristics of the soil sample in the initial five cycles in the typical stage are described in Fig. 3.24, including cyclic creep curve, pore water pressure curve, and hysteresis curve. Figure 3.24a is the cyclic creep curve. The figure shows that under the initial loading, cyclic creep curve is surrounded by peak creep curve and valley creep curve, moving to the upper right with the approximately sinusoidal method gradually. The three creep characteristic curves have the same trends, as the strain increases rapidly while the rate of strain is in decay. From the analysis of the four curves, we can observe that at the beginning of loading, the cyclic creep varies and increases within a certain cyclic strain amplitude range in the base of the basic creep. Figure 3.24b shows the variation of the pore water pressure in the soil. In the initial stage, the pore pressure varies and increases with the approximately sinusoidal method. But compared with the creep curve, the waveform is slightly irregular. Figure 3.24c is the hysteresis curve of the deviator stress-strain relationship. The visco-plasticity can be reacted in hysteresis curve. The curve in the figure moves to the right and tends to be intensive rapidly, indicating the good visco-plasticity of the soil. The soil will accumulate a certain amount of plastic strain in each loading cycle. But the accumulation amount is gradually reduced, i.e., plastic strain gradually accumulates, but the accumulation rate decays rapidly.

Figure 3.25 describes the characteristics of the soil sample in the final five cycles in the typical stage, including cyclic creep curve, pore water pressure curve, and deviator stress-strain hysteresis curve. The cyclic creep curve in Fig. 3.25a shows the axial strain changes in sinusoidal waveform, and the basic creep curve is parallel to the horizontal direction. The three creep curves change to be nearly parallel and horizontal, reflecting that the creep is almost stable, the basic creep does not increase any more, and the accumulating rate of plastic strain in the soil sample decreased nearly to be zero. The primary creep varies and increases within a certain cyclic strain amplitude range in the base of the basic creep. Figure 3.25b is the pore pressure curve in the final state. It is going to be horizontal and stable regardless of a little irregular. Figure 3.25c shows us a superposition of the deviator stress-strain curve. It proves again that with the time of loading cycles increasing, the soil grains gradually tend to be dense and plastic deformation tends to mitigate. Consequently, the hysteresis hoops are going to be overlapped, but not move to the right.

The comparative analysis of Figs. 3.24 and 3.25 shows the cyclic creep of saturated mucky clay presents the elastic and plastic characteristics simultaneously. With the increase of loading times, the elasticity tends to be stable and visible, while the plasticity goes weak. The primary creep varies and increases within a certain cyclic strain amplitude range in the base of the basic creep. That strain amplitude is not immutable. It tends to be an approximate fixed value with the growth of loading time. Throughout the process, the cyclic creep experienced the stages of rapid growth—slow growth—tending to be stable in turn. Due to the impact

3.5 Creep Behavior of Soil Under Cyclic Loading (Tang et al. 2010)

Fig. 3.24 Mechanical characteristics of the soil sample in the initial five cycles: (**a**) creep strain; (**b**) pore water pressure; (**c**) axial stress strain

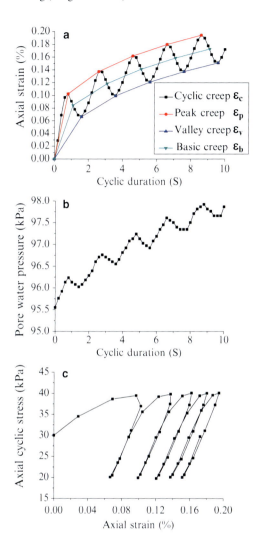

of the test instruments and soil quality, the cyclic characteristics of the pore water pressure is slightly irregular compared with the creep. But it still shows the similar characteristics with the creep on the whole. Soil's deviator stress-strain curve moves to the right gradually, and the hysteresis loop goes intensive gradually and eventually coincident, which is the evidence of the elasticity and plasticity of the cyclic creep.

3.5.4.3 Composition of Creep

According to Fig. 3.25a, peak creep curve, basic creep curve, and valley creep curve change to be nearly parallel and with the same distance. Define variable ε_e as the distance between the parallels and ε_b as the average of peak and valley value.

Fig. 3.25 Mechanical characteristics of the soil sample in the last five cycles: (**a**) creep strain; (**b**) pore water pressure; (**c**) axial stress strain

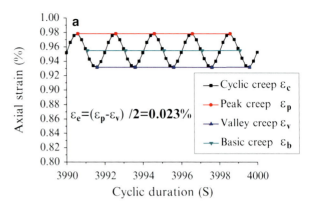

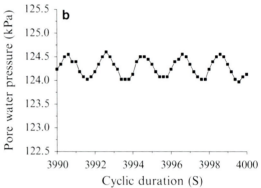

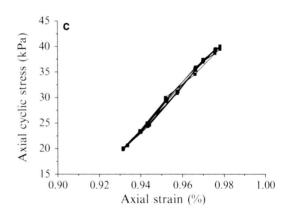

$$\varepsilon_e = \frac{(\varepsilon_p - \varepsilon_v)}{2} \tag{3.8}$$

$$\varepsilon_b = \frac{(\varepsilon_p + \varepsilon_v)}{2} \tag{3.9}$$

3.5 Creep Behavior of Soil Under Cyclic Loading (Tang et al. 2010)

Thus,

$$\varepsilon_p = \varepsilon_b + \varepsilon_e \tag{3.10}$$

$$\varepsilon_v = \varepsilon_b - \varepsilon_e \tag{3.11}$$

Equations 3.10 and 3.11 indicate that the total cyclic creep process of mucky clay changes in a certain amplitude (ε_e) on the basis of basic creep (ε_b). The generation and recovery of ε_e is an elastic reversible changing process. So we define ε_e as the reversible elastic strain and equation (3.8) is the calculation formula for reversible elastic strain of every loading cycle.

The incremental tendency of basic creep curve in Fig. 3.25 implies that some irreversible plastic strain accumulation could be produced during each loading cycle. The basic creep curve can be obtained by connecting these plastic strain value points. Thus, the nature of basic creep is the accumulated plastic strain. This is the basis of the study on long-term rheological behavior of soils.

The above analysis shows that cyclic creep process of the saturated mucky clay is composed of two strains: One is relatively constant reversible elastic strain, and the other is progressively attenuating accumulated plastic strain, which represents elastic and plastic characteristics of the soil, respectively. Under the long-term cyclic loads, the reversible elastic strain tends to be stable and the nature of basic creep is the accumulated plastic strain.

3.5.5 The Change of Reversible Elastic Strain of Saturated Clay

3.5.5.1 Variation of Reversible Elastic Strain of Soil Under Different Stress Level

Step loading is a widely used method in cyclic creep test. Stress level of each step could be controlled by two parameters: CSA (cyclic stress amplitude) and CSR (cyclic stress ratio). CSR is defined using the equation proposed by Professor Seed in 1996:

$$\text{CSR} = \frac{\text{CSA}}{2\sigma'} = \frac{3\text{CSA}}{2(\sigma_1 + \sigma_2 + \sigma_3)} \quad (\sigma_2 = \sigma_3) \tag{3.12}$$

The same analytical method of C sample in the typical stage and $\varepsilon_e = (\varepsilon_p - \varepsilon_v)/2$ was applied to analyze and calculate the reversible elastic strain values of C and D samples in each cyclic loading step under different strain level. C sample experienced 7 cyclic loading stages under a frequency of 0.5 Hz. It presented failure trend when the cyclic strain amplitude reaches 35 kPa, and 7 reversible elastic strain

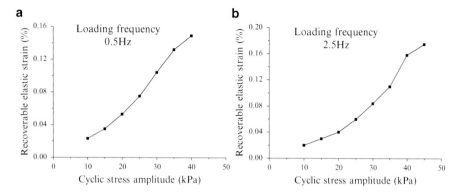

Fig. 3.26 Recoverable elastic strain versus cyclic stress amplitude curves with loading frequency: (a) $f = 0.5$ Hz; (b) $f = 2.5$ Hz

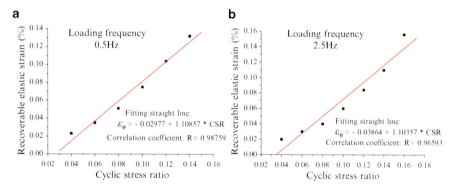

Fig. 3.27 Fitting lines between recoverable elastic strain and cyclic stress ratio with loading frequency: (a) 0.5 Hz; (b) 2.5 Hz

values were obtained. Similarly, D sample experienced 8 cyclic loading stages under frequencies of 2.5 Hz. It tended to be a failure at 40 kPa, and 8 reversible elastic strain values were obtained.

Figure 3.26 shows relation of reversible elastic strain with different cyclic stress amplitudes, (a) for sample C and (b) for sample D. It can be obviously seen that reversible elastic strain increases with stress amplitude. Before the sample becomes a failure, its reversible elastic strain becomes larger with the cyclic stress amplitude increasing. Figure 3.27 was plotted by considering cyclic stress ratio CSR of the stress level and recoverable elastic strain. The last point value of recoverable elastic strain was ignored because the sample was close to failure in this case. It could be seen from this figure that there is a good linear relationship between recoverable elastic strain and CSR before soil becomes a failure. Huang Hongwei and Zhu et al. (2005) also got similar conclusion.

3.5 Creep Behavior of Soil Under Cyclic Loading (Tang et al. 2010)   83

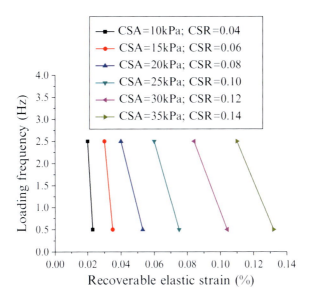

Fig. 3.28 Loading frequency versus recoverable elastic strain under different cyclic stress ratios

3.5.5.2 The Influence of Loading Frequency to Recoverable Elastic Strain of Soil

The relation between loading frequency and recoverable elastic strain under different cyclic stress ratios is shown in Fig. 3.28, from which conclusions can be drawn that the value of recoverable elastic strain of saturated mucky clay is influenced by loading frequency. Under the same stress level, lower frequency load resulted in a larger recoverable elastic strain. And as it is seen in the figure, the slope of lines changes from nearly vertical to sloping and the gradient is greater and greater with the stress level increasing. Furthermore, we can know from the contrast on the linear fitting formulas of the two figures in the last section that the loading frequency has no significant influence on the linear relationship between reversible elastic strain and cyclic stress ratio.

3.5.6 Variation of Accumulated Plastic Strain of Saturated Clay

3.5.6.1 Threshold Cyclic Stress Ratio of Soil

According to the results of step cyclic loading test, the accumulated plastic strain curve of saturated mucky clay can be plotted. Figure 3.29 reflects the accumulation of plastic strain at a loading frequency of 0.5 Hz in Fig. 3.29a of C sample and

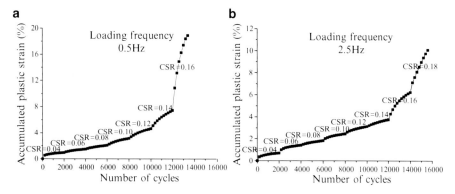

Fig. 3.29 Accumulated plastic strain versus number of cycles with loading frequency: (a) $f = 0.5$ Hz; (b) $f = 2.5$ Hz

2.5 Hz in Fig. 3.29b of D sample, respectively. The trend of the two curves implies that there are two threshold cyclic stress ratio values under step cyclic loading, thus dividing the variation of the accumulated plastic strain into gradual stabilization stage, rapid growth stage, and instantaneous failure stage.

The first threshold value exists between the gradual stabilization stage and rapid growth stage. It can distinguish whether the soil's strain will tend to stabilize under the long-term loading. It can be defined as stability cyclic stress ratio. If the stress level of cyclic loading is less than that value, then soil's long-term accumulated strain will tend to stabilize. On the contrary, if the stress level of cyclic loading is greater than that value, soil will be destroyed because of the too large accumulated strain in the long or short term under different stress level. Figure 3.29a and b shows that the accumulated plastic curves are nearly horizontal. Values of the two CSRs are both 0.08 and the corresponding cyclic stress amplitudes are both 20 kPa with a frequency of 0.5 and 2.5 Hz.

The second threshold value exists between the rapid growth stage and instantaneous failure stage. It can distinguish whether the soil will be destroyed instantaneously with the rapid increase of accumulated strain. It can be defined as the failure cyclic stress ratio. If the stress level of cyclic loading is greater than that value, the soil will be destroyed because of the too large deformation within a very short time. On the contrary, if the stress level of cyclic loading is less than that critical value, the soil structure will be destroyed due to too large accumulated deformation under the long-term loading. The trend of the two curves is obvious. The curves rise with a certain slope in the rapid growth stage. Through the failure cyclic stress ratio, the curves suddenly produce a turning point and at last come to almost vertical. Figure 3.29a shows that the value of CSR is 0.14 and the corresponding cyclic stress amplitude is 35 kPa with a frequency of 0.5 Hz. Figure 3.29b shows that the value of CSR is 0.16 and the corresponding cyclic stress amplitude is 40 kPa with a frequency of 2.5 Hz.

3.5 Creep Behavior of Soil Under Cyclic Loading (Tang et al. 2010)

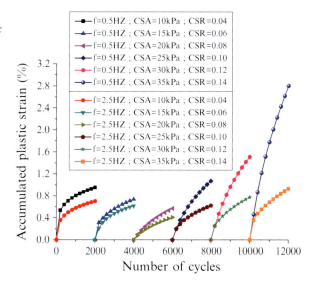

Fig. 3.30 Comparative curves of accumulated plastic strains with different loading frequencies (**a**) $f = 0.5$ Hz; (**b**) $f = 2.5$ Hz

The comparison of the two threshold cyclic stress ratios reflects that the frequency of cyclic loading has no significant influence on the soil's stability cyclic stress ratio, while it has an influence on the soil's failure cyclic stress ratio, that is, failure cyclic stress ratio gets greater under cyclic loading with higher frequency.

3.5.6.2 Variation of Accumulated Plastic Strain by Loading Frequency

The comparison of accumulated plastic strain under each step loading and different loading frequencies is depicted in Fig. 3.30. The 12 curves can be divided into 6 groups according to the standard of the same stress levels and different loading frequencies. The accumulated strain curves are all above the contrast curves with a loading frequency of 0.5 Hz. It shows that under the same loading level, the accumulated plastic strains are greater under cyclic loading with lower frequency. And a significant trend of the curves under the same loading frequency can be detected that the former half segment of the curve is gradually decaying while the latter part turns into accelerated growth, which also proved the turning point of stability cyclic stress ratio.

3.5.7 Variation of the Residual Pore Water Pressure

Because of the viscous properties of saturated mucky clay, the pore water pressure generated in clay cannot dissipate quickly under cyclic loading, leading to the formation of residual pore water pressure. Figure 3.31 describes the variation of

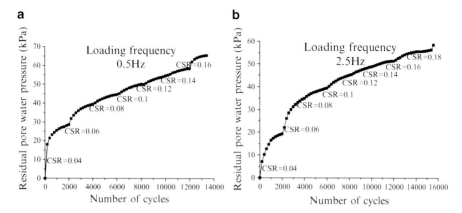

Fig. 3.31 Residual pore water pressure versus number of cycles with loading frequency: (**a**) $f = 0.5$ Hz; (**b**) $f = 2.5$ Hz

residual pore water pressure in the soil samples under step cyclic loading. It can be seen that the pore water pressure had a rapid growth initially and then maintained uniform growth after the growth rate decayed to a stable value. When the stress level was close to the failure cyclic stress ratio, the development of pore water pressure became irregular. In addition, by comparison, the figure implies that it generates larger residual pore water pressure in mucky clay in Shanghai when subjected to cyclic loading of lower frequency.

This rule is similar with the aforementioned accumulated plastic strain. Under the same stress levels, it generates larger accumulated plastic strain when subjected to the loading of lower frequency.

3.5.8 Summaries

In order to solve the engineering geologic problems caused by long-term subway operation loading, the cyclic creep property of mucky clay in Shanghai was investigated through cyclic triaxial tests under step loading. The results and relevant conclusions of experimental study are summarized and drawn below:

1. Two threshold cyclic stress ratios, the stability cyclic stress ratio and the failure cyclic stress ratio, exist and divide the cyclic creep development into three stages, including the gradual stabilization stage when stress level is below stability CSR, rapid growth stage when stress level is between stability CSR and failure cyclic stress ratio, and instantaneous failure stage when stress level is above failure cyclic stress ratio.
2. The cyclic creep of the mucky clay in Shanghai is composed of two parts, that is, relatively constant recoverable elastic strain and progressively attenuating

accumulated plastic strain. They are obviously detected from the three characteristic creep curves. Especially when the stress level is below the failure cyclic stress ratio, there is a linear relation between the recoverable elastic strain and the cyclic stress ratio.
3. It generates greater recoverable elastic strain, accumulated plastic strain, and residual pore water pressure in mucky clay in Shanghai, when subjected to the cyclic loading of lower frequency under the same stress level in the frequency section of 0.5–2.5 Hz. The loading frequency has a significant influence on the failure cyclic stress ratio, but not on the stability cyclic stress ratio. And there is larger failure cyclic stress ratio with the loading frequency increasing.

3.6 Dynamic Stress-Strain Relationships

The stress-strain relation of soils depends on the initial conditions (void ratio, saturation, pore structure, and effective stress) and is also relevant to the external loading mode and boundary conditions. Some scholars pointed out that the strain softening type of stress-strain relation was suitable for most soils under low stress level. But whether the saturated soft clay behaves similarly under cyclic vibration loads, the discussion is presented in this part as follows. The strain softening effects in saturated soft clay under cyclic loads have not been much documented. Matasovic and Vucetic (1992, 1998) established the relation between a softening parameter t and cyclic strain. The soil stain softening was reflected by the variation of t. Vucetic and Dobry (1998) explored the influences of over-consolidation ratio and plastic index on soil softening. Yao and Nie (1994) modified the previously proposed softening parameter t and made it more applicable to avoid the zero value when vibration loading time was close to infinite. Wang and Yao (1996) proposed a simulation model of boundary element, in which he used boundary radius or maximum elastoplastic modulus at the initial loading to describe the isotropic softening in soils. Shen (1993) described the soil softening based on the view of damage. All these have great guidance on study of softening under cyclic loads.

In this section, all tests were performed in CKC cyclic triaxial system as well. The specific experimental conditions were still detailed in Sect. 3.2.1. Nevertheless, the dynamic loading was conducted under consolidation ratio of 0.46 on mucky clay. The consolidation loads were applied in stages and less than 0.005 mm in the lateral deformation was the stable criterion in each loading stage. The dynamic stress σ_d and strain ε_d results were plotted in Fig. 3.32 under various confining pressure σ_3. Relations of $1/E_d - \varepsilon_d$, $E_d - \varepsilon_d$ were presented in Figs. 3.33 and 3.34 as well for further discussion.

In Fig. 3.21 the relations of stress-strain were derived from the tests under a loading frequency of 2 Hz and a vibration time of 60 min. From Fig.3.22, $1/E_d$ was roughly proportional to ε_d as expressed in formula $1/E_d = a + b\varepsilon_d$. Through transposition $E_d - \varepsilon_d$ relation can be obtained as $E_d = 1/(a + b\varepsilon_d)$. In conjunction with $E_d = \sigma_d/\varepsilon_d$, $\sigma_d - \varepsilon_d$ relation can be formed in formula as $\sigma_d = \varepsilon_d/(a + b\varepsilon_d)$.

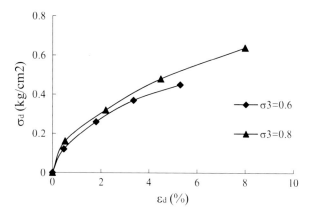

Fig. 3.32 Curves dynamic stress and strain of soft silt clay

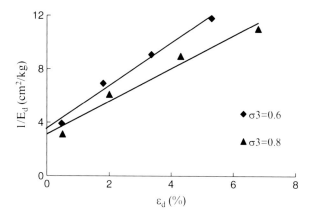

Fig. 3.33 Curves of $1/E_d - \varepsilon_d$

Hence, it made sense that the relation of dynamic stress strain can also be expressed as Kondner hyperbolic curve. The fitted equations were presented below by regression analysis.

When $\sigma_3 = 0.6$ kg/cm^3,

$$\frac{1}{E_d} = 1.6\varepsilon_d + 0.036 \tag{3.13}$$

When $\sigma_3 = 0.8$ kg/cm^3,

$$\frac{1}{E_d} = 1.2\varepsilon_d + 0.031 \tag{3.14}$$

3.6 Dynamic Stress-Strain Relationships

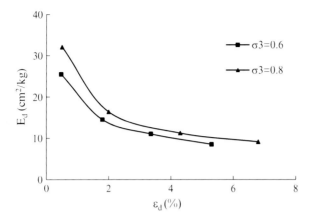

Fig. 3.34 Curves of $E_d - \varepsilon_d$

Table 3.9 Relevant parameter values

σ_3 (kg/cm^2)	a (cm^2/kg)	b	E_{od} (kg/cm^2)	G_{od} (kg/cm^2)
0.6	0.036	1.6	28	10.4
0.8	0.031	1.2	32	11.9

Fig. 3.35 Curves of $1/G_d \sim \gamma_d$

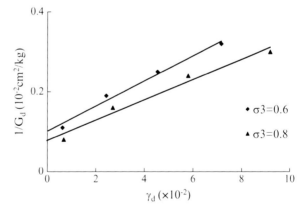

According to Eqs. (3.8) and (3.9), parameters a, b and maximum shear modulus E_{od} were obtained, and the corresponding maximum dynamic shear modulus can be calculated as well by $G_{od} = E_{od}/[2(1 + \mu)]$, where μ was the Poisson ratio of 0.35 in mucky clay. All relevant parameter values were presented in Table 3.9.

In addition, the parameter of γ_d and G_d can be derived by relations of $\gamma_d = \varepsilon_d(1 + \mu)$, $G_d = E_d/[2(1 + \mu)]$. The results were shown in Figs. 3.35 and 3.36.

Experimental results indicated that the dynamic elastic modulus E_d and dynamic shear modulus G_d both varied with different confining pressures and dynamic stress levels. Under the same consolidation ratio, they increased with confining pressure and declined with dynamic stress; and these behaviors were much conspicuous when

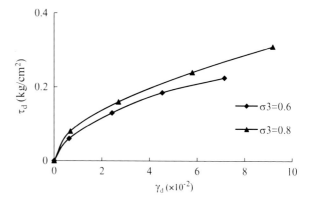

Fig. 3.36 Curves of $\tau_d \sim \gamma_d$

the strain was small. This was accordant with the properties in deformation that it rose fast at the beginning of loading and trended to stabilization gradually. In the initial period of loading, an instantaneous deformation occurred, and even if the value was small, dynamic shear modulus changed a lot. After a certain loading duration, the response in soils was not as obvious as in the previous stage. Creep effects played the main role in soil deformation. Hence, all the curves had the similar characteristics, that is, firstly ramped fast and then slowed down gradually.

3.7 Effective Principal Stress Variation

The soil deformation is greatly relevant to the variation of effective stress. Hence, from the view of effective stress theory, it is a best way to explore the deformation characteristics of soils under reciprocated loads.

3.7.1 Experimental Introduction

In this part the GDS bender element system was used, which enables easy measurement of the maximum shear modulus of a soil at small strains in a triaxial cell. The testing preparation and procedures were the same with previous specific description. The experimental design can be referred to Table 3.1.

The S- and P-wave testing can be performed in this GDS bender element system. The working principle is shown in Fig. 3.37. The signal transmitter and receiver of GDS titanium bender elements are the main function components. They are distributed in two ends of the testing specimen. The signal is transmitted out by the transmitter, traveling through the specimen, and finally received by the receiver element. The transmitter and receiver are both connected with high-speed data acquisition system and dedicated software. The graphical user interface (GUI) is presented in Fig. 3.38. It is characterized with properties of easily control, real-time testing and recording, and high reliability.

3.7 Effective Principal Stress Variation

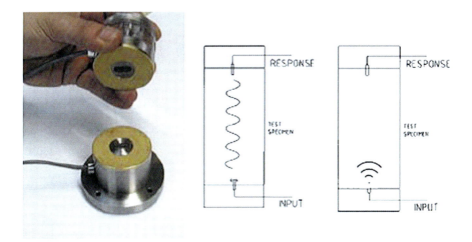

Fig. 3.37 Working principles of GDS BES

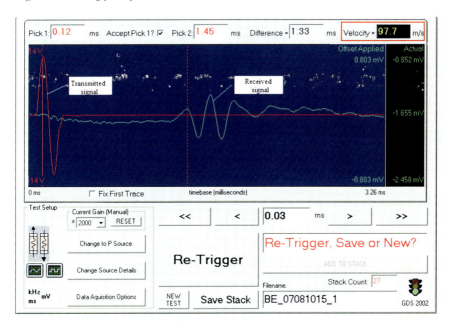

Fig. 3.38 Working interface of GDS BES

After the full consolidation, firstly the original shear wave velocity of the testing specimen was measured before it was applied to the vibration loads. Then, each shear wave velocity testing was conducted respectively after vibration loading cycles of 500, 1,000, 2,000, 3,000, 4,000, 5,000, 7,500, 10,000, 15,000, 20,000, etc., until the end of loading.

Fig. 3.39 Deformation of specimen after test

3.7.2 The Variation of Effective Primary Stress with Loading Time

Once the cyclic triaxial testing was finished, the specimen was taken out and investigated, as shown in Fig. 3.39. It showed that even if the stress condition was relatively low, conspicuous deformation arose in the surface and in the middle some obvious heave occurred.

Figure 3.40a–c presented the variations of effective primary stress (i.e., effective axial stress in this testing condition) with vibration loading cycles corresponding to field depth of 8.5, 11.5, and 13.5 m.

Under different confining stress, the effective axial stress in each case all sharply declined negative proportionally to loading time in the previous period. Then gradually the curve deflection slowed down. Through a long-time decaying, the effective axial stress approached a stable value. From the curves, the attenuation of the effective axial stress can be divided into three stages (take Fig. 3.40a as an example to describe as sharply declining stage AB, transition stage BC, and stabilization stage CD).

3.7.2.1 Sharply Declining Stage

Once the cyclic loads were applied, the effective primary stress sharply decreased by 15 % during a short time. Since the dynamic stress σ_d was much smaller than the

3.7 Effective Principal Stress Variation

Fig. 3.40 Curves of effective principal stress versus vibration cycles number

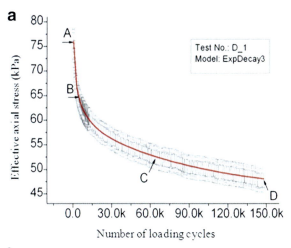
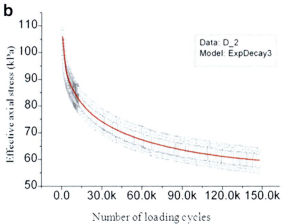
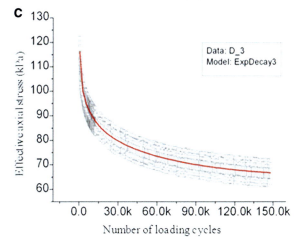

Table 3.10 Variation parameters of effective principal stress

Test no.	σ_0	N_0	A_1	A_2	A_3
830D1	64.6706	−8,309.15168	35.81288	15,345.9494	27.07312
826D2	57.08206	−5,948.48918	1,392.18537	24.29459	27.86796
828D3	45.1514	−3,258.68999	11.67635	151.05315	15.63458

Test no.	δ_1	δ_2	δ_3	R^2
830D1	9,397.74181	1,236.73351	60,968.50944	0.8944
826D2	1,281.42006	10,667.03066	66,321.89615	0.91126
828D3	8,431.29202	1,247.12497	90,354.52702	0.91768

primary stress, great excess pore water pressure must be generated from the view of effective stress theory. In the end of this stage, the falling rate of the effective primary stress started to slow down.

3.7.2.2 Transition Stage

This stage was a long-time transition process. The decaying of the effective primary stress continuously developed, while the absolute value was still large. It indicated that the excess pore water pressure rose up quickly but the growth rate was gradually slowed down.

3.7.2.3 Stabilization Stage

In this stage, the variation of the effective primary stress was not much conspicuous and gradually stabilized into a specific value, which was termed as the critical effective primary stress under this vibration loading condition. Results in these three group tests were all approximate to 60 % (66, 57, and 60 %).

In addition, as the stress condition became larger, the sharply declining stage lasted a shorter time. This implied that the excess pore water pressure growth rate at this stage increased with stress condition.

Discussion of the relation of effective primary stress and loading time facilitated the study on deformation mechanism. Combining the own properties of Shanghai soft clay and engineering project background, a logistic model curve was proposed based on regression analysis on the data:

$$\sigma' = \sigma_0 + A_1 \cdot \exp\left(-\frac{N-N_0}{\delta_1}\right) + A_2 \cdot \exp\left(-\frac{N-N_0}{\delta_2}\right) + A_3 \cdot \exp\left(-\frac{N-N_0}{\delta_3}\right) \quad (3.15)$$

where σ' was effective primary stress, N was loading cycles, and σ_0, N_0, A_1, A_2, A_3, δ_1, δ_2, δ_3 were regression parameters. The specific values were all presented in Table 3.10.

3.7.3 The Variation of Shear Wave Velocity

The variation of shear wave velocity in each specimen was collected by GDS BES and presented in Fig. 3.41.

From the curves, it can be found that during the previous sharply declining stage of the effective primary stress, the excess pore water pressure ramped quickly (while the axial strain was in the resilience stage); the shear wave velocity decreased a lot due to the rise in damping. When the effective primary stress attenuation slowed down (at this time, the axial strain resumed to rise), the falling rate of shear wave velocity slowed down as well. Then, effective primary stress got into stabilization stage; the shear wave velocity also decreased slowly and stably. The decrease in shear wave velocity during this stage was mainly induced by soil structure damage (breakage and rearrangement of soil particle skeleton). By virtue of regression analysis on shear wave velocity data, the ExpDecay2 model can be fitted well, as shown in Eq. (3.16):

$$v_s = v_{s0} + A \cdot \exp\left(-\frac{N - N_0}{\xi_1}\right) + B \cdot \exp\left(-\frac{N - N_0}{\xi_2}\right) \quad (3.16)$$

where v_s was the shear wave velocity, N was loading cycles, and $v_{s0}, N_0, A, B, \xi_1, \xi_2$ were regression parameters. The specific values were all presented in Table 3.11.

3.8 Mechanism Analysis

Researchers regarded that the structure properties normally existed in soils. Most naturally deposited normally consolidated clay has exclusive structure behaviors. And the undisturbed structural clay is much similar to the nonhomogeneous structural gravel cemented aggregates (Shen 1993). The cyclic triaxial test results indicated visible resilience phenomenon, akin to the shear dilatancy in sands.

The energy generated by subway vibration in operation transmitted into the surrounding soil by track and lining structure. Since the pore water is much more sensitive than soil particles, in the previous loading stage, the energy was mainly absorbed by pore water resulting in sharp rise in pore water pressure. Correspondingly, the effective stress greatly decreased in a short time. Elastic stress release occurred in the aggregates. As the vibration lasted, the effective stress was gradually close to stabilization, and the transferred energy started to be consumed by soil particle aggregate. The weak bonds among soil aggregates were broken down and damaged. A lot of micro-cracks arose. Until the energy accumulated in soil structure to some extent, all these cracks were well connected throughout the whole soil particle skeleton, severely segmenting the soil structure and forming shear bands. The particles within these zones are further pulverized.

It is postulated that the total deformation increment is composed of two parts. One is induced by the increase of effective stress, which can be calculated by existed elastoplastic model; the other aspect is the particle breakage (such as

Fig. 3.41 Curves of shear wave velocity versus time

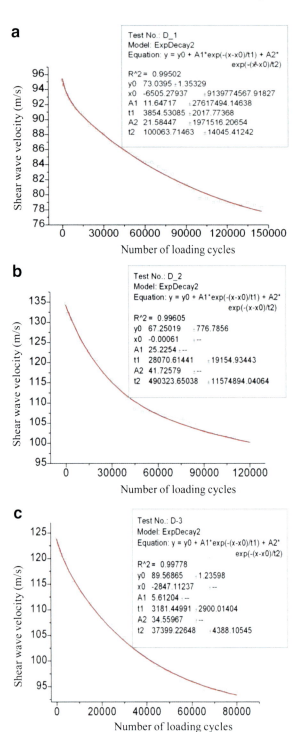

Table 3.11 Parameters of decline curves of shear wave velocity

Test no.	v_{s0}	A	B	N_0	ξ_1	ξ_2	R^2
830D1	73.0395	11.64717	21.58447	−6,505.27937	3,854.53085	100,063.71	0.99502
826D2	67.25019	25.2254	41.72579	−0.00061	28,070.61441	490,323.65	0.99605
828D3	89.56865	5.61204	34.55967	−2,847.11237	3,181.44991	37,399.22	0.99778

creep) of soil particle skeleton. The particle damage can be generated purely by compressive stress, and the shear stress aggravates the damage. Assuming f is the yielding surface of particle sliding and g is the damage function of sol particles, the corresponding strain increment can be expressed as shown in Eq. (3.17) according to orthogonal flow rule:

$$\{\Delta\varepsilon\} = [C]\{\Delta\sigma\} + A_p\{\partial f/\partial\sigma\}\Delta f + A_d\{\partial g/\partial\sigma\}\Delta g \quad (3.17)$$

where A_p is plasticity coefficient and A_d is the damage coefficient. Recommended by Shen, f and g can be formed as below:

$$f = \sigma_m/[1 - (\eta/\alpha)^n]; \quad g = \sigma_m/[1 - (\eta/\beta)^n] \quad (3.18)$$

where $\sigma_m = (\sigma_1 + \sigma_2 + \sigma_3)/3$ and α, β, η are the needed parameters, which can be determined by secondary mercury intrusion porosimetry.

3.9 Chapter Summary

Through a series of laboratory tests under cyclic loads on saturated soft clay in Shanghai and Nanjing, the pore water pressure, soil dynamic strength, dynamic stress-strain relation, and dynamic elastic modulus were discussed deeply, taking vibration loading frequency, amplitude, and vibration loading cycles into account. The dynamic properties and microstructure deformation damage mechanism were analyzed. Some important conclusions were drawn as below:

1. The development of pore water pressure can be divided into three stages: fast ramping stage, slowly growth stage, and stabilization stage. The logistic model was proposed to fit the pore water pressure growth curves and coefficients of correlation that were all above 0.99. At the fast ramping stage, the growth rate was not always a constant. The pore water pressure increased sharply, slowed down in a short time, and then got into stabilization. After regression analysis, the growth rate of pore water pressure obeyed in the ExpDecay2-type curve. In addition, the excess pore water pressure declined abruptly once the vibration loading ceased and finally recovered at a marginal higher value compared to static pore water (hydrostatic) pressure.
2. The deformation characteristics of saturated soft clay under cyclic loads were mainly influenced by cyclic stress ratio, confining pressure, vibration loading frequency, and vibration cycle number. There existed a critical cyclic stress ratio,

relevant to the soil properties, applied load form and consolidation pressure, etc. When the applied dynamic load was smaller than the critical cyclic stress, the dynamic stress increased with vibration loading time. While the growth rate and the amplitude gradually fell into a stable constant. In contrast, the deformation increased fast, and finally reached its peak value. Under low loading frequency, the influence of frequency was not very conspicuous and can be neglected. While in the high-frequency circumstance, the soil behaved more elastic properties. The main reason may be that when high-frequency vibration load is applied, the pore water pressure could not dissipate in time, and with longer loading cycle duration, the pore water pressure dissipated fully and the creep effects on soil particles were more obvious. The soil characteristics are more plasticity and viscosity.

3. As for the subway tunnel, the deformation in the soil deep around the interior wall of the subway tunnel was relatively small, while the axial strain curve of soils just beneath the subway tunnel was in a very short resilience stage and got into the plastic deformation stage very quickly, resulting in large lateral deformation. Hence, even if there may be no large deformation occurring around the subway tunnel during a long operation time, significant differential settlement is probably to arise as time lasts.

References

Andersen KH, Rosenbrand WF, Brown SF et al (1980) Cyclic and static laboratory tests on Drammen clay. J Geotech Eng Div ASCE 106:499–529
Boulanger RW, Idriss IM (2007) Evaluation of cyclic softening in silts and clays. J Geotech Geoenviron Eng ASCE 133:641–652. doi:10.1061/1090-0241
Chai JC, Miura N (2002) Traffic load induced permanent deformation of road on soft subsoil. J Geotech Geoenviron Eng ASCE 128:907–916. doi:10.1061/1090-0241
Chan CK (1981) An electropneumatic cyclic loading system. Geotech Test J 4:183
Gong QM, Zhou SH, Wang BL (2004) Variation of pore pressure and liquefaction of soil in metro. Chin J Geotech Eng 26(2):290–299 (in Chinese)
Gouy G (1910) J Phys (France) 9:457
Guan ZC, Jiang YJ, Tanabashi Y (2009) Rheological parameter estimation for the prediction of long-term deformations in conventional tunneling. Tunn Undergr Space Technol 24:250–259
Guo Y, Luan MT, He Y, Xu CS (2005) Effect of variation of principal stress orientation during cyclic loading on undrained dynamic behavior of saturated loose sands. Chin J Geotech Eng 27(4):403–409 (in Chinese)
Huang MS, Li JJ, Li XZ (2006) Cumulative deformation behavior of soft clay in cyclic undrained tests. Chin J Geotech Eng 28(7):891–895 (in Chinese)
Hyodo M, Yasuhara K, Hirao K (1992) Prediction of clay behavior in undrained and partially drained cyclic tests. Soils Found 32(4):117–127
Jiang J (2000) Analysis for the behavior of clay and sand cored composite specimens under cyclic loading [D]. Zhejiang University, Hangzhou
Jiang J, Chen LZ (2001) One dimensional settlement of soft clay under long-term cyclic loading. Chin J Geotech Eng 23(3):366–369
Li LY, Cui J, Jing LP, Du XL (2005) Study on liquefaction of saturated silty soil under cyclic loading. Rock Soil Mech 26(10):1663–1666 (in Chinese)
Lo KY (1969) The pore pressure-strain relationship of normally consolidated undisturbed clays. Can Geotech J 6(383):412
Matasovic N, Vucetic M (1992) Pore pressure model for cyclic straining of clay. Soils Found 32(3):156–173

References

Matasovic N, Vucetic M (1995) Generalized cyclic degradation pore pressure generation model for clays. J Geotech Eng 121(1):33–42

Matsui T (1980) Cyclic stress strain history and shear characteristics of clay. J Geotech Eng 106(10):1101–1120

Meng QS, Wang R, Chen Z (2004) Pore water pressure mode of oozy soft clay under impact loading. Rock Soil Mech 25(7):1017–1022 (in Chinese)

Miura N, Fujikawa K, Sakai A, Hara K (1995) Field measurement of settlement in Saga airport highway subjected to traffic load. Tsuchi-to-kiso 43(6):49–51

Monismith CL, Ogawa N, Freeme CR (1975) Permanent deformation characteristics of subgrade soils due to repeated loading. Transportation Research Board, Washington, DC, pp 1–17

Moses GG, Rao SN, Rao PN (2003) Undrained strength behaviour of a cemented marine clay under monotonic and cyclic loading. Ocean Eng 30:1765–1789

Parr GB (1972) Some aspects of the behavior of London clay under repeated loading. Dissertation, University of Nottingham, Nottingham

Seed HB, Chan CK (1961) Effect of duration of stress application on soil deformation under repeated loading. In: Proceedings of the 5th international Congress on soil mechanics and foundations, vol 1. Dunod, Paris, pp 341–345

Shao LT, Hong S, Zhen WF (2006) Experimental study on deformation of saturated sand under cyclic pore water pressure. Chin J Geotech Eng 28(4):428–431 (in Chinese)

Shen ZJ (1993) An elasto-plastic damage model for cemented clays. Chin J Geotech Eng 15(3):21–28 (in Chinese)

Skempton AW (1954) The pore water pressure coefficient A and B. Geotechnique 26:317–330

Tang YQ, Huang Y, Ye WM et al (2003) Critical dynamic stress ratio and dynamic strain analysis of soil around the tunnel under subway train loading. Chin J Rock Mech Eng 22(9):1566–1570

Tang YQ, Wang YL, Huang Y et al (2004) Dynamic strength and dynamic stress-strain Relation of silt soil under traffic loading. J Tongji Univ Nat Sci 32(6):701–704

Tang YQ, Zhou J, Liu S, Yang P, Wang JX (2010) Test on cyclic creep behavior of mucky clay in Shanghai under step cyclic loading. Environ Earth Sci 63(321):327

Vucetic M, Dobry R (1998) Degradation of marine clays under cyclic loading. J Geotech Eng 114(2):133–149

Wang R, Liu JH (2000) Monitoring and studying of Shanghai subway's longitudinal deformation during its long time running. Undergr Eng Tunn 4:6–11

Wang JH, Yao ML (1996) Elastoplastic simulation of the cyclic undrained behavior of soft clays. Chin J Geotech Eng 18(3):11–18 (In Chinese)

Wu MZ, Zhou HB, Chen ZC (1998) Test analysis of degradation behavior of saturated soft clay after cyclic loading. J Tongji Univ 26(3):274–278 (in Chinese)

Yao ML, Nie SL (1994) A model for calculating deformation of saturated soft clay. J Hydraul Eng 7(51):55 (in Chinese)

Yılmaz MT, Pekcan O, Bakır BS (2004) Undrained cyclic shear and deformation behavior of silt–clay mixtures of Adapazarı. Turk Soil Dyn Earthq Eng 24:497–507

Zeng QJ, Zhou B, Gong XN, Bai NF (2001) Growth and dissipation of pore water pressure in saturated soft clay under impact load. Chin J Rock Mech Eng 20(1):1137–1141 (in Chinese)

Zeng CN, Liu HL, Feng TG, Gao YF (2005) Test study on pore water pressure mode of saturated silt. Rock Soil Mech 26(12):1963–1966 (in Chinese)

Zhang KL, Tao ZY (1994) The prediction of pore pressure of saturated clay under cyclic loading. Rock Soil Mech 15(3):9–17 (in Chinese)

Zhang R, Tu YJ, Fei WP, Zhao ZH (2006) Effect of vibration frequency on dynamic properties of saturated cohesive soil. Rock Soil Mech 27(5):699–704 (in Chinese)

Zhou J, Tu HQ (1996) A model for predicting the cyclic behavior of soft clay. Rock Soil Mech 17(1):54–60

Zhou HL, Wang XH (2002) Study on the pore water pressure of saturated sand in dynamic triaxial test. J China Railw Soc 24(6):93–98 (in Chinese)

Zhu DF, Huang HW, Yin JH (2005) Cyclic creep behavior of saturated soft clay. Chin J Geotech Eng 27(9):1060–1064

Chapter 4
Research of Microstructure

4.1 Introduction

The macroscopic engineering properties of soil are greatly controlled by the state and behavior of microstructure, which makes the study of microstructure crucial. However, the soil microstructure is so complicated that it can be hardly simulated or represented by any homogeneous continuum model.

As early as 1925, Terzaghi proposed the concept of "honeycomb structure" for the soil structure. Subsequently, Goldschmidt (1926) and Casagrande (1932) were also aware of the significance of the microstructure in soil mechanics and made further development of Terzaghi's structural mode. *Micropedology* written by Kubiena established a significant foundation for the formation and development of microstructure theory. However, because of the limitation of research methods, the research of soil microstructure soon stepped into a slow development period. Since the 1950s, the application of optical microscopy, polarizing microscopy, and X-ray diffraction provided great convenience for the study on microstructure. Meanwhile, the occurrence of dipping method provided a great method to prepare high-quality thin sections of soil, which promoted the microstructure study significantly as well. More people were inspired to study the microstructure. The first international soil micromorphology research conference was initiated by Volkenrode, Free, and Altermuller. The success of this conference pushed the soil microstructure research to a new stage. More studies were concentrated on the processes of structure formation from the perspective of the sedimentary environment. Brewer (1964) began to pay attention to the integrity of the structural elements, especially to the characteristics of orientation distribution of structure unit. In addition, he also put forward a new soil micromorphology analysis system based on the structural description system proposed by Kubiena; furthermore, he first integrally proposed the concept of "fabric." Silverman (1960), Chen (1957), and Lambe (1953) applied X-ray diffraction and polarizing microscopy to conduct quantitative tests of the orientation arrangement of mineral sections (clay mineral sections). At the end

Fig. 4.1 Cross section through the center of the pore channels (pore throat d and ventral pore d')

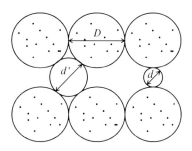

of 1960s, the application of scanning electron microscopy, electron probe, X-ray diffraction, polarizing microscopy, and a large number of advanced techniques brought the soil morphology research to the next level. According to a large number of research results, it revealed that fixed and neat combination modes as crystal structure or molecular structure hardly exist in clay, its modes have considerable randomness, and the particle arrangements are very complicated. Under longtime cyclic loading, the deformation of soil, which is related to the soil strength, increases with time growth. In other words, soil deformation and displacement are determined by microstructure variation (such as the relative displacement and movement between structure elements). The structure elements maintain dynamic balance during soil deformation process. The soil microstructure has unique characteristics both in time and spatial domains. Indeed, scanning electron microscopy (SEM) was used to be the main method for the study on clay microscopic morphology (Bai and Zhou 2001; Chen et al. 2000; Shan et al. 2004; Zhou and Mu 2005). For instance, Deng et al. (2005) made an analysis on microstructure of recently deposited collapsible loess in Yongdeng, Gansu province in China; Shan et al. (2004) studied on the microstructure transformation of silt obtained from the Yellow River delta affected by hydrodynamic effects; Wang et al. (2005) studied on the microstructure of soft soil. Tang et al. (2005) studied on the microscopic characteristic of saturated soft clay through scanning electron microscopy test (SEM). The main content of the microstructure study includes the shape and size of soil elements, contact state and connection form between the basic elements, the pore shape and size, and the mechanical properties in the three stages of soil deformation. In addition, the soil deformation mechanism was explained by comparing undisturbed soil samples before and after cyclic triaxial test. However, SEM method has defects which cannot be ignored: (1) it is difficult to accurately quantify the SEM images; (2) the SEM images are 2-D and not accurate enough to reflect the soil structure difference or its changes under different engineering environment; (3) because of the contradiction between magnification and soil sample representativeness, the magnification could not be unrestrainedly improved; therefore, generally the maximum pore diameter in SEM images is in the micron level; and (4) according to the seepage theory, it is the pore throats instead of pore size that influence water penetration (Fig. 4.1).

Limited by the disadvantages in SEM method, its research results are basically applied in qualitative analysis. However, benefiting from the development of computer technology, image processing technology, and the combination of

4.1 Introduction

interdisciplinary knowledge and technical methods (such as mercury intrusion porosimetry (MIP), GIS, Videolab image analysis), several structure parameters can be obtained currently. The improvements are primarily reflected in the following three aspects: (1) extracting the quantization parameters of soil structure by processing the soil microstructure images with the help of image processing technology, (2) applying fractal theory developed in the 1970s to make quantitative research on the soil microstructure, and (3) using GIS-integrated information technology to acquire quantization parameters of structure elements. The research group led by Tovey and Osipov has made significant improvement in computer quantitative processing of clayey soil microstructure morphology; and the team of Bazant got a new breakthrough in clayed soil micromechanics; Wu (1991) developed a quantitative analysis system of soil microstructure images which can describe several clay structure parameters (structure entropy, average particle diameter of structure elements, average shape) and analyzed the relationship between the structure changes and special clay physical mechanical properties; Shi applied D/MAXIIIA automatic fabric goniometer and Videolab image analysis system in the quantitative research of clayey soil microstructure orientation. Shi (1997; Shi and Jiang 2001) preliminarily established observation technology system and quantitative analysis system for clayey soil microstructure and firstly proposed frequency distribution function of aggregates and simple quantitative analysis method of microstructure. Hu (1995) indicated that soil microstructure is constituted by four structure parameters: particle morphology (particle size, particle shape, and surface roughness), particle arrangement (grain orientation and particle size distribution), porosity (pore size and pore size distribution), and particle contact state (contact form and connectivity). Hu also systematically studied on the abundant information and quantitative parameters of these four aspects and put forward an entropy quantitative expression of soil size fraction. Through combining computer technology and mathematical statistics, Wang et al. (2001) proposed three kinds of quantitative indicators and quantitative evaluation methods for evaluating the microstructure orientation of clayey soil. Gong (2002) studied on the soft clay particles and aggregate composition, pore size distribution, microstructure, pore solution, and cation exchange capacity. He also studied the influence of pore variation caused by consolidation and artificial recharge on soil properties. From a physical-chemical view, the influence of micro features on consolidation deformation and land subsidence was also discussed. Li and Bao (2003) calculated the porosity with microscopic structure images based on the grayscale algorithm in 3D dimensions.

The fractal theory was introduced to soil microstructure research since the early 1990s. Xiao and Akilov (1991) proposed to use porosity fractal dimension to characterize the clay nonuniformity. Xie (1992) studied on the fractal characteristic of pores and particles of geological materials and proposed a pore fractal measurement method. Hu (1995) obtained the quantitative results of the particle size distribution, surface fractal dimension, and fractal dimension distribution of pores and contact zones and analyzed the relationship between the engineering properties and these parameters based on Hausdorff fractal dimension calculation method. Wang and Wang (2000) used mercury intrusion porosimetry method to measure the soil pore and proposed the distinction of clayey soil pore classification with the actual distribution of soil pore structure combined with the view of scale-free interval of fractal geometry theory. Using a data extraction technology combined with image processing technology with geographical information system (GIS) software, Wang et al. (2004) calculated the fractal dimension of soil particle morphology and discussed the determination of the threshold value during the image processing. In addition, he also demonstrated the clayey soil microstructure fractal characteristic and analyzed the relationship between the fractal dimension and microstructure of the soil based on the microstructure image analysis of expansive soil samples. Wang et al. (2008) studied on the 2D and 3D calculation methods for soil porosity by applying SEM images based on the area and volume calculation methods provided by GIS.

Although many scholars all over the world study on the microstructure and have achieved gratifying achievements, rare research has focused on the change rules of saturated soft soil microstructure under subway loading. A large number of published research results demonstrate that clay does not have fixed or neat modes as crystal structure or molecular structure, and the particle arrangements are very complicated and random. The property of soft soil varies in different regions. So the microscopic structure of the soil has its unique characteristics both in time and spatial domains. Under vehicle loading, the soil deformation and displacement are mainly determined by internal structure; and the structure elements keep dynamic equilibrium during the process of soil deformation (Hu 2002; Chen 2001). Although a lot of research achievements about the microscopic structure have been made both in China and abroad, they are mostly in the stage of theory exploration; research achievements of microstructure are hardly applied to solve specific engineering problems.

Therefore, this research focused on quantitative research of soil samples before and after vibration using AUTOSCAN 60 mercury intrusion porosimetry (MIP) produced by the USA. The study is aimed to (1) find the variation of pore size distribution, pore quantity, and other characteristic parameters before and after subjected to vibration loading, (2) visually and qualitatively analyze the SEM images, and (3) finally establish their relationship with corresponding macroscopic mechanical properties.

4.2 Qualitative Analysis

The study purpose of soil microstructure includes two aspects: one is to understand the nature of soil engineering properties; the second is to predict variations of geotechnical engineering characteristics. The characteristics of clay's microstructure can be described from three aspects: (1) morphological characteristics, such as size, shape, surface characteristics, and relevant quantitative proportion; (2) geometry characteristics, such as spatial configuration of structure elements; and (3) energy characteristics, such as structural connection type and total structure energy. At present, the microstructure research progress mainly includes the following four parts: (1) open structure and clay sensitivity, (2) relationship between fabric differences and soil deformation or strength, (3) structure and soil expansibility, and (4) collapsibility of loess. The first three are mainly concerned in this study.

1. Open Structure and Clay Sensitivity
   Many scholars have done structure research for a long time, and the early structure theories were mainly based on the research of clay sensitivity. In terms of medium-sensitivity soil and extremely active soil, the basic arrangement of soil aggregates is flocculated structure composed by fabric and ladder-shaped fabric or visible granular texture. For all kinds of sensitive soil, clay particle matrix is visible. In addition, the contribution of soil sensitivity to soil structure characteristics is mainly due to the appearance of relatively open microstructure formed by irregular aggregates, namely, the unstable connection between soil aggregates.
2. Relationship Between Fabric Differences and Soil Deformation or Strength
   Research demonstrates that soil fabric has obvious effect on the deformation characteristics. As for dispersed structure and flocculated structure, the former needs much smaller deviator stress to obtain equivalent deformation than the latter in the low stress environment. However, when the stress rises to a certain value, the flocculated structure begins to destroy and the deviator stress reduces dramatically. Finally the variations are consistent with each other.

   The strength of the soil decreases with the augment of the orientation degree; the shear deformation can destroy soil particles and aggregates. With the occurrence of slip plane, laminar or strip mineral will prefer to slip along its long axis, leading to the decrease of strength until reaching a residual value. According to the research results obtained by Sergeyev and Osipov, soil with a certain orientation demonstrates large shear resistance but small cohesion, whereas it is opposite in the soil with random arrangement. When the shear direction is perpendicular to the soil particle orientation, the soil's strength is 30 % larger than the one under a parallel shearing. Cohesion reduces with the increase of orientation degree of structural aggregates on the shear plane. The $\tau - f(\sigma)$ relationship is not linear in some occasions. The shear stress of clay with far coagulation contact type decreases with the increase of vertical load, while the $\tau - f(\sigma)$ relationship can maintain linear in the clay with phase contact type (Sergeyev et al. 1980).
3. Structure and Soil Expansibility
   The expansibility of soil is mainly due to the reduction of effective stress caused by unloading or water percolating; and microstructure plays a significant role in the generation and development of soil expansion. Osipov opined that the expansibility of soil was closely related with clay particle matrix aggregation and structural contact was the main control factor. Soils with far coagulation contact type and phase contact type have no expansibility, while the soil with ion-electrostatic contact has significant expansibility. In addition, soil of near

coagulation contact type and mixed contact shows weak expansibility. When soil natural structure is destroyed, its expansibility is mainly decided by soil mineral composition and physical condition.

Although research about microstructure has made great achievements in the past 70 years, the development speed is quite slow as many problems and insufficiencies exist. The main problems include the following three aspects: (1) the quantification of soil structure need to be solved; (2) the existing quantitative information extraction methods need to be popularized; meanwhile, new technical methods are urged to be found, especially for solving test problems about the structure connection and morphology elements. Furthermore, convenient test technology of bonding structure needs more development. (3) Coupling problem of quantitative structure and mechanics model needs further study.

Scanning electron microscopy (SEM) technology has been widely used in soil microstructure research and acquired abundant SEM photos during microstructure analysis of soil samples. Different scales of the SEM photos reflect the spatial distribution of soil particle aggregates and morphology characteristics of connection on different layers of structure. The application of SEM for qualitative analysis is very mature and widely used.

This chapter primarily includes the qualitative analysis of soil microstructure changes under dynamic vehicle loading and the soil deformation mechanism. Soil samples in this research mainly employed mucky clay and fine silty sand in Nanjing Subway and saturated soft clay in Shanghai Subway.

4.2.1 Research Method

SEM firstly scans the sample surface through using fine focusing electron beam point by point. Then, various physical signals are generated by the interaction of electron beams and samples. Then, the signals are transferred to modulation signal after receiving by detector and amplification. At last, images reflecting characteristics of the sample surface are shown on the screen. SEM is an efficient analysis tool for the study of sample surface with a great number of advantages, such as deep field, strong stereo sense, large image magnification range, continuous adjusting, high resolution, large sample room space, and efficient sample preparation.

The accelerating voltage needed by SEM which is generally about 1–30 kV is much lower than that of transmission electron microscopy, and it can be chosen according to the nature of samples properly. The common acceleration voltage is about 20 kV. The image magnification of SEM can be continuously adjusted within a certain range (a few times to hundreds of thousands times) as the magnification is the ratio of lateral length displayed on the screen and the actual length. The electronic optical system differing from the transmission electron microscopy (SEM) is just applied to provide scanning beam as the excitation source of all physical signals produced by soil sample. The most common signals used by SEM are secondary electron signals and backscattered electron signals; the former is used to display surface morphology contrast, while the other one is used to display atomic number contrast.

The basic structure of SEM can be divided into six parts which are electronic optical system, scanning system, signal detection amplification system, image display and record system, vacuum system, and power supply and control system. The electron gun of SEM emits electron beam which is focused to a hairline beam on the sample surface after electric field acceleration and two or three electromagnetic lens. This fine electron beam under the function of the double deflection coil on the top of the lens scans the surface of the sample and excites all kinds of signals, such as secondary electron, backscattered electron, absorb electronics, X-ray, auger electron, and cathodoluminescence, after the interaction of accelerated electrons and the sample compositions. These signals are converted into a brighter or darker portion of the image in an ordered and proportional way at a given point (x, y), and then an image is formed as the electron beam moves across the sample area after scanning and amplification. At last, different kinds of characteristic images of the sample surface can be observed on the screen.

Generally, geotechnical engineering focuses on the study of mechanics characteristics (deformation and strength) of geological materials and their influencing factors, especially on the structure, fabric, and relationship between deformation and strength mechanism of geological materials. SEM provides an access to study the composition, arrangement, and connection of soil and expands the research perspective of engineering geology. Zhang (1986) provided material for engineering geological mechanics evaluation using SEM to study soil microstructure and morphology. Du (1990) discussed the relationship between microstructure characteristics and permeability of loess using SEM to observe the microstructure and characteristics.

4.2.2 Preparation of Samples

1. The seal of undisturbed soil samples were opened carefully in order to avoid disturbance. Then, according to the test requirements, the samples were divided into two parts as one part is used for dynamic triaxial test under cyclic loading while the other part is used for microstructure research of undisturbed soil.
2. According to the test requirements, part of sample with certain water content was chosen and breaks by hand without using tools or equipment. As for these samples with large viscosity, knife was applied to cut a fine seam in order to break easily. If the soil sample had smooth joint plane inside, it would be easy to break down through that smooth surface. The microstructure morphology observed would be the microstructure characteristics of the smooth surface part rather than the real soil mass structure, so choosing a typical observation surface during the preparation of samples is very important.

3. The soil sample was made in certain size according to the direction of observation, with the help of magnifying glass; samples with flat surface were selected for test research.
4. The specimens were put in an oven and dried with a certain temperature (<100 °C) and time.
5. Then the dried specimens were put into the spray apparatus to spray a thin gold coat, which was manipulated to increase contrast between pores and soil units so that regions of light pixels represented soil units and dark pixels represented pores.

4.2.3 Basic Characteristics of the Soil Samples

4.2.3.1 Mucky Clay in Nanjing Subway Area

The mucky clay of Nanjing Subway is buried at the depth of 11 m; it is the recently deposited soil with loose structure and low length under the environment of slowly flowing water and microorganisms. The mineral compositions are quartz, feldspar, mica, a lot of clay minerals (especially montmorillonite), organic matter, and a small amount of water-soluble salt mineral. Due to the characteristics above, the mucky clay of Nanjing Subway has excellent hydrophilism, and the main structure is honeycomb structure. The main engineering performance has the following features:

It has high porosity ratio and saturated water content. The natural water content is over the liquid limit. Void ratio is generally 1.21, liquid limit is 44 %, degree of saturation is about 95 %, and natural water content is 50 %. The permeability is very poor due to the interbeded thin silt layer. The vertical permeability is weaker than horizontal. It has high compressibility with compression index of 5 kPa^{-1} which increases with the growth of water content. This low consolidation degree is a result of massive hydrophilic minerals in soil and short deposition time. The shear strength is low relating to the loading rate and drainage conditions. Under the undrained quick shear test, $\varphi = 0$; c is generally less than 0.25 kPa; under drained consolidation quick shear condition, φ is generally $10°-15°$ and c is generally 25 kPa.

4.2.3.2 Saturated Soft Clay in Shanghai Subway Area

The saturated soft clay was formed in the environment with slowly flowing water and influenced by microorganisms. It is the recent alluvial soil with loose structure and low length. Besides quartz, feldspar, and mica, a lot of clay minerals (especially montmorillonite), organic matter, and a small amount of water-soluble salt mineral are also included. The soft clay nearby Shanghai Subway has excellent hydrophilism in honeycomb structure. The main engineering performance has the following features:

1. High Porosity and Water Content
   Void ratio is generally 0.8–1.3; liquid limit is 30–60 %; natural water content is 40–60 %. Undisturbed soil is in soft plastic state.
2. Weak Permeability
   The permeability coefficient of saturated soft clay is generally 8.64×10^{-6}–8.64×10^{-4} m/d. Interbeded by thin fine sand layer, the vertical permeability is often smaller than the horizontal.
3. High Compressibility
   The compressibility coefficient $a_{0.1-0.2}$ is generally 0.7–1.5 MPa^{-1}, and it gets larger with an increase of water content.
4. Low Shear Strength
   The shear strength is related to the loading speed and drainage conditions. Under drained consolidation quick shear condition, φ value is generally 0° and c value is less than 20 kPa, while under the undrained, φ is generally 611° and c is 9.0–11.0 kPa.

4.2.4 Qualitative Analysis of Observed Result

Compared with the SEM images of Shanghai saturated soft clay, the images of Nanjing mucky clay are not so excellent both in quality and quantity because of relatively earlier stage technology, insufficient time, and less advanced SEM device. However, in order to integrate the research and show the development of microscopic study, this part is divided into three sections concentrated on microscopic qualitative analysis according to different soil types, especially on the results of Shanghai saturated soft clay.

4.2.4.1 Mucky Clay of Nanjing Subway

Due to the special formation environment of mucky clay, as well as the corresponding particle and mineral compositions, there is a hydrated water film on the surface of clay particle which connects interparticles together and provides cohesive force in mucky clay.

The photos (Fig. 4.2) provided by SEM show that the microstructure of Nanjing Subway mucky clay is commonly honeycomb structure, which is formed by the deposition of the fine soil fraction in water and disintegration under a superimposed load. During the depositing process, an edge-to-face contact flocculent structure is gradually composed as porous loose structure with chain bodies. Soil structure is fluffy just after depositing as a flocculated structure. Then, because of the changes of pressure environment, especially the increase of shear forces, the edge-to-face contacts in soil are turned into face-to-face contacts gradually, and the

Fig. 4.2 Microstructure of undisturbed samples under different multiples (**a**) 638×; (**b**) 5,557×; (**c**) 6,887×

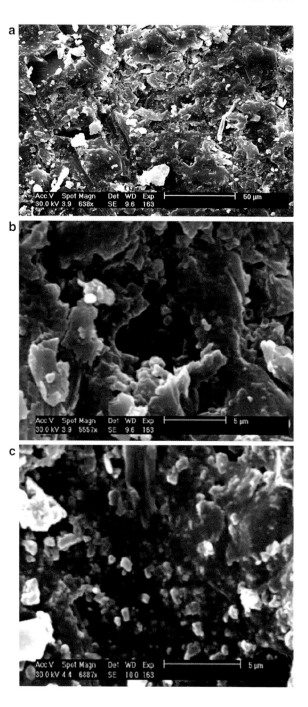

4.2 Qualitative Analysis

honeycomb structure is formed. This loose dispersed structure is gradually oriented into laminated structure and fabric structure. It possesses characteristics of low shear strength, high compressibility, and low permeability.

The vibration-subsidence refers to the subsidence of mucky clay layer under subway traffic vibration. The dynamic triaxial test results (Fig. 3.10) in previous chapter indicate that on one hand, under a certain condition, the permanent deformation value gradually tends to be stable with the increase of time when the applied dynamic stress is less than the critical dynamic stress value. The fastest deformation rate is in the initial stage of loading, while with the increase of vibration loading cycles, deformation rate also decreases to zero gradually. On the other hand, when applied dynamic stress is larger than the critical dynamic stress value, the soil finally gets into failure with the increase of time. In addition, according to the dynamic triaxial test results in previous chapter, the deformation is larger and easy to be destroyed in one cycle with larger cell pressure; larger lateral pressure coefficient k in consolidation stage also leads to the destruction of soil. Combined with microstructure analysis, the following mechanism can be made as follows.

The mucky clay primarily has honeycomb structure with high porosity, high saturated water content, and certain cohesion as a consequence of a hydrated water film on particle surface. When the applied dynamic stress is less than the critical dynamic stress, the induced microstructure changes can be divided into three stages:

1. Consolidation stage: During the soil sample consolidation stage, microstructure adjustment (pure disintegrating or polymerization in grain particles) is obvious. The reason is that the soil samples out of the ground have experienced an unloading process and its structure "relaxation" because of the stress release, and the applied pressure for consolidation is mainly used to make structure back to the natural conditions of equilibrium state and thus the structure adjustment is significant.
2. Microstructure rearrangement stage: During this stage, with the subway dynamic load, the structure changes when the cohesion of soil could not balance external forces and it behaves in several aspects: decrease of particle size, compression of pores, decrease of the porosity and a certain change of particle arrangement orientation, etc. Actually, this stage is a transitional phase during which the old structure system formed under natural condition is gradually broken down to the adjustment of the changing pressure environment. This phase is the main deformation stage.
3. Structure-solidifying stage: During this stage, if the dynamic stress is smaller than the critical dynamic stress, soil can achieve a consolidated balance structure to bear the load, and the permanent deformation value can be stable with soil structure adjustment. However, when the dynamic stress is greater than the critical dynamic stress, the soil cannot resist forces through continuous structure

adjustment. At this time, the connection strength and structure between soil particles are damaged, even broken down; and the soil water pressure increases dramatically and is conveyed to the water in soil pores by the particle contact point. At last the soil structure gets to destruction.

When the soil lateral pressure coefficient is small, namely, the cell pressure applied for consolidation is small. The soil has suffered from great initial shear stress with large initial shear displacement. In terms of the soil microstructure, the shear forces applied on soil lead to a new equilibrium of soil particle size distribution, particle orientation arrangement, and pore sizes. After the structure-solidifying stage, the bonding strength between soil particles is strengthened along with the improvement of shearing strength of soil. Therefore, from the macroscopic view, the soil permanent deformation is smaller under the influence of dynamic loads. This is the reason why the deformation is smaller with lower cell pressure.

Figure 4.2 shows the microscopic structure of undisturbed mucky clay under different magnifications. Figure 4.3 indicates the microstructure with stable deformation under the condition of $\sigma_3 = 0.6$ kg/cm^2, lateral pressure coefficient $k = 0.46$, consolidation time equals to 6 h, dynamic stress ratio $R = 0.3$, and vibration cycles $N = 105$. In comparison of these two figures, we can see that the soil has certain changes of the grain orientation, surface roughness degree, particle size, and shape after vibration. Figure 4.2a shows that the arrangement of particle orientation is chaotic. And Fig. 4.3a shows that the particle arrangement tends to be perpendicular to the horizontal plane. Figure 4.3b can also reveal that tendency. The reason for this phenomenon is that the spatial distribution of soil particle is optimized to adapt to the new pressure environment and to achieve new consolidated balance structure due to the vertical cyclic load. Thus, soil structure reflects good orientation after vibration which is consistent with the decrease of the shearing strength of soil. In the shearing process, for the soil whose particle orientations are disordered, part of energy will be consumed for particle rotation; while for the soil whose particle orientations are ordered, no such energy will be consumed, which means the same shear stress can induce larger displacement. Thus, the undisturbed soil has higher shear strength to reach a shear failure. In addition, according to the two figures, the particle size increases and particle size distribution is more even with lower particle surface roughness degree. Generally speaking, if the roughness of particle surface is larger, namely, the particle shape is vertically irregular, it is more probable that the connections between particles are point contacts, which means the soil has a structure with poor stability and can be easily compressed.

According to the soil microstructure changes in the consolidation stage, the microstructure rearrangement stage, and the structure-solidifying stage, it can be understood why the compression performance of soil are improved under train-induced vibration.

According to the dynamic triaxial test result analysis in previous chapter, the deformation of silty sand has changed sharply under dynamic loads at first. Then the

4.2 Qualitative Analysis

Fig. 4.3 Microstructure of samples after vibration under different multiples (**a**) 580×; (**b**) 2,319×

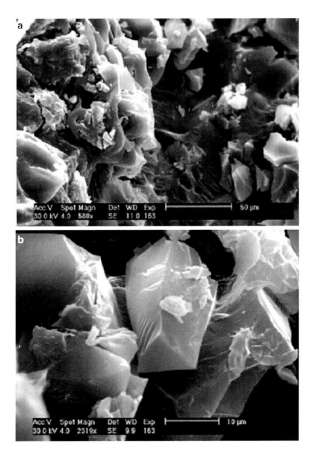

deformation tends to be stable along with the growth of the vibration time. If applied dynamic stress is less than the critical dynamic stress, permanent deformation value tends to be stable with the increase of the vibration cycles. Otherwise, the soil sample is destroyed after a period of vibration. According to this phenomenon, combined with the soil structure in meso-level and microlevel, the following mechanism analysis can be made:

At the beginning of cyclic loading tests, large instantaneous deformation is appeared in silty sand because of the weak attraction between sand particles and the high water content of soil as well. The main reason is that when the particles are compressed, the pore volume is decreased; part of the stress transfers to the excess pore water pressure and soil skeleton strength is reduced; finally the residual deformation is generated. When the applied dynamic stress is less than the critical dynamic stress, soil particles are able to adapt to the changing environment through continuous adjustment and the deformation decays; this is a structure rearrangement stage. Finally, when vibration time continues to grow, soil structure adjustment is not obvious and the changes of most structural parameters tend to be tiny. The new structure system has been basically established and the structural elements only make appropriate adjustments to ensure more consolidated balance structure with further effect of pressure. In this stage, the permanent deformation is basically stable. But if applied dynamic stress is greater than the critical dynamic stress, soil structure still cannot adapt to the new pressure environment with a period of adjustment as the vibration cycles increase. During this process, pore water pressure keeps rising with effective stress decreasing, and eventually the soil loses its strength, like the liquefaction in silty sand layer.

4.2.4.2 Saturated Soft Clay of Shanghai Subway

First of all, the soil particle contact state, shape, size, connection form, and pore shape and size are observed and described by SEM. And based on these observations, further research is made about the soil microstructure deformation mechanism.

The mineral composition and structure of clay determines its physical, chemical, and mechanical properties. Due to the fine grain size, the polarized light microscopy method is difficult to identify them. During this experimental program, X-ray diffraction (XRD) analysis, SEM, and energy spectrum analysis method are used to identify the mineral composition.

X-Ray Diffraction Analysis Principle

The X-ray wavelength is almost equal to the distance between crystal mineral internal atoms (or ions); wave diffraction phenomenon is produced with a beam of X-rays through the clay mineral lattice of different crystal face. As clay mineral has bedding structure, diffraction phenomenon happens according to the Bragg formula when X-ray goes through the crystal face.

Test Method

Clay soil samples were taken from subway station of Shimen Road, and the sampling depth was 16.00–16.30 m. The purpose of the analysis is to acknowledge the main content of clay minerals in this clay soil. Clay minerals were extracted from sample by conventional method to make oriented section. All the tests were conducted in X-ray diffraction laboratory of Marine Department in Tongji University. And all the testing includes the natural air drying section, saturation section by ethylene glycol (EG), and section by 550 °C heating treatment.

4.2 Qualitative Analysis

Testing conditions are as follows:

Diffractometer type: PW1710 based	Tube anode: Cu
Generator tension [kV]: 40	Generator current[mV]: 20
Wavelength alpha1 [Å]: 1.54056	Wavelength alpha2[Å]: 1.54439
Intensity ratio (alpha2/alpha1): 0.500	Divergence silt: Automatic
Irradiated length [mm]: 12	Receiving silt: 0.1
Spinner: On	Monochromator used: Yes
Start angle [°2θ]: 3.000	End angle[°2θ]: 36.000
Step size [°2θ]: 0.010	Time per step[S]: 0.500
Type of scan: Continuous	Intensities converted to: Fixed

Analysis Results

X-ray diffraction curve of clay minerals is presented in Fig. 4.4.

X-ray diffraction analysis which is the most reliable method for clay mineral identification can study crystal structure of clay minerals. According to the test results, the clay minerals of this sample are mainly illite (49.3 %), montmorillonite (8 %), kaolinite (14.9 %), and chlorite (27.8 %).

Clay minerals have adsorption characteristics. The water capacity and ion exchange performance of clay minerals are improved after adsorbing organic matter, while particle space increases and connection force decreases. The growth of soil particle dispersion leads to large soil plasticity, lower permeability and strength, and higher compressibility. Montmorillonite has the strongest adsorption ability of organic among different kinds of clay minerals.

Generally, microstructure refers to the microscopic particle skeleton and the connection between the particles, as shown in Fig. 4.5. Most of the research methods regard soil microstructure as the main factor to determine the soil mechanics characteristic. However, according to related literatures, few detailed evidence can be seen to support that the shear strength is influenced by particle composition or inter-aggregate connection force. Therefore, the research program focuses on quantifying the micro fabric of saturated clay and evaluating their influences on the shear strength. Clay microstructure includes aggregate shape, aggregate structure, pore structure, and connectivity. And all these characteristics can be expressed by nine different parameters.

The Shape and Size of Microstructure Basic Element

The microstructure basic element refers to solid unit with obvious physical boundaries which can bear certain force under a certain magnification scale. The characteristic of microstructure basic element includes particle composition, size, shape, and surface characteristics. Generally speaking, the element can be a single mineral particle (namely, "particle grain") and can also be multiple mineral particle aggregation

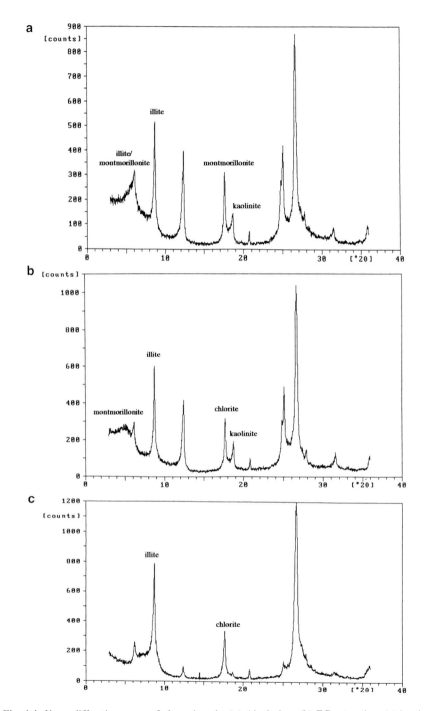

Fig. 4.4 X-ray diffraction curve of clay minerals: (**a**) Air drying; (**b**) EG saturation; (**c**) heating treatment (cps is the unit of diffraction intensity. It refers to the electron number received by counter per seconds)

4.2 Qualitative Analysis

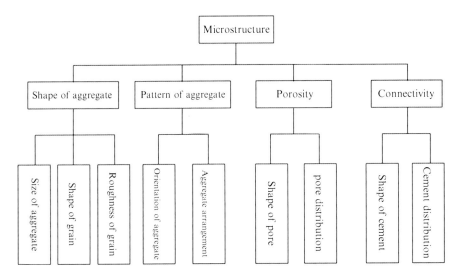

Fig. 4.5 Microstructure of clay skeleton

(namely, "soil aggregation"). Normally the microstructure basic elements can be divided into two levels—primary and secondary. The primary element has strong original cohesive force which is micro coagulated and difficult to separate. And the secondary element is commonly a laminar or platy aggregation or granular aggregate formed by primary micro coagulation.

This chapter chooses SEM images before and after vibration (Fig. 4.6). It can be seen that there are little changes of the grain size and shape before and after vibration, and the arrangement of particles seems to be more regular after vibration. In contrast, the size and shape of basic elements on horizontal and vertical surface of undisturbed saturated soft clay sample and vibration sample with cyclic triaxial loading failure can be observed clearly. Comparing each two of them (a–b, c–d), we can see that the soil particles with certain orientation are broken after vibration. This is because after vibration, soil particles have rearrangement process. Quantitative analysis will be discussed in Sect. 4.3.

The Contact State of Basic Elements

Commonly, the arrangements of particles are in three different ways shown in Fig. 4.7, namely, face-to-face contact, edge-to-edge contact, and edge-to-face contact. It can be seen in SEM images of the vertical surface section (Fig. 4.8) that almost all the clay particles are connected around in face-to-face, edge-to-edge, or edge-to-face contact types. The clay particles look like thin platy and the aggregates are flocculated look like flowers and feathers. In contrast, the horizontal surface which is difficult to distinguish the edge and face contact has more granular

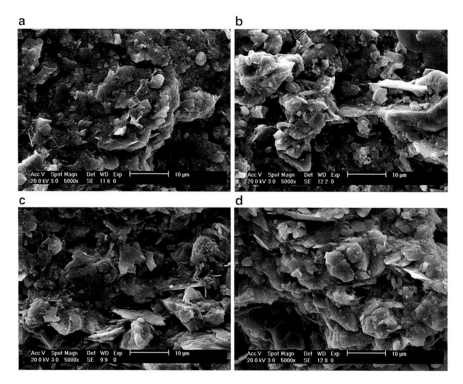

Fig. 4.6 Microstructure of samples after vibration (×5,000): (**a**) horizontal surface of undisturbed sample; (**b**) horizontal surface of sample after vibration; (**c**) vertical surface of undisturbed sample; (**d**) vertical surface of sample after vibration

Fig. 4.7 Arrangement of mineral particles: (**a**) face-to-face; (**b**) edge-to-edge; (**c**) edge-to-face

coagulation, and it is mostly in direct contact and mosaic contact state while some particles are connected by bonding.

The Connection Form of Basic Elements

Clay structure connection mainly refers to interaction between structure basic elements and connection properties. Normally, the structure connection can be clas-

4.2 Qualitative Analysis

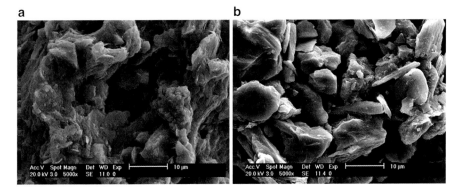

Fig. 4.8 Contact state of particles: (**a**) vertical surface and (**b**) horizontal surface

sified by two methods: (1) according to differences of inter-aggregate composition, the connection types are divided into five categories: no connection (or mosaic contact), ice connection, capillary water connection, bound water connection, and bonding connection; (2) regarding only inter-aggregate force can reflect the nature of the structure connection, and classification is based on the inter-aggregate distance and force. And the structure connection mainly has three basic types which are coagulation, transitional, and phase type. In addition, the coagulation type can also be divided into two types which are far coagulation and near coagulation type depending on strength of the contact force.

According to the figures above, some elements contact with each other directly because of certain substances (electrical charge, water film, adhesive film, etc.), and some elements are connected by bonding material obviously. High-magnification SEM images show that mucky clay contains two different types of microcrystal: One is salt crystal which adheres to the surface of particles, or fills in pores, or forms larger aggregates. The other kind of crystal is pyrite microcrystalline. And a single microcrystalline is in octahedron or cube shape. They generally gather together forming spherical aggregates in the pore of clay or on the particle surface. Mucky clay also contains a certain number of silt, biological fabrics, and diatomaceous earth (shown in Fig. 4.9). They have great influence on soil compressibility, permeability, and porosity.

Shape, Size, and Classification of Pore

The soil pore type, size, and shape play an important role in the soil compressibility, permeability, and consolidation properties. The isolated pore (intra-aggregate pore) of mucky clay with poor connectivity mainly exists in the secondary aggregates and has influence on the compressibility of soil but not the permeability. The inter-aggregate pore exists between granular aggregates and clay aggregates. The quantity of inter-aggregate pore is very large, and it plays an important role in soil porosity,

Fig. 4.9 SEM images of saturated soft clay soil in Shanghai: (**a**) salt crystals; (**b**) aerial pore structure; (**c**) biological fabrics

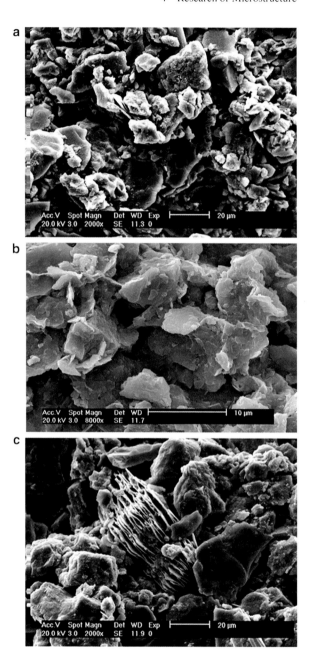

compressibility, permeability, and consolidation because of its good connectivity. Intra-aggregate pore mainly exists in the clay aggregates, and the pore is small with large dispersion and poor connectivity. It mainly influences the compressibility of soil.

4.2 Qualitative Analysis

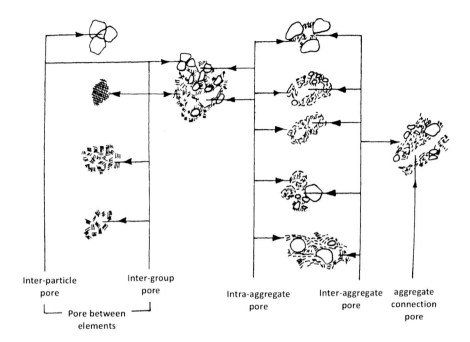

Fig. 4.10 Classification of pore structure in clay soil

An important indicator of pore characteristics is pore quantity, namely, the porosity. However, the parameter is related to the size, shape, orientation, and arrangement of structure elements. Thus, it is inadequate to evaluate the soil engineering properties through the soil porosity only. Generally speaking, the characteristics of soil pore structure include the pore size, shape, number, and connectivity since studies show that they have important influence on soil engineering geological and hydrogeological properties.

By SEM, the pore in aggregates and fine-grained elements can be seen, but the pore in microparticles is hardly visible. In addition, the pore among basic elements and large pore connecting several elements can also be seen, and the size and shape of pore depend on the boundary conditions and arrangement situation of soil elements. With the help of image analyzer and other technical means, the pore size can be measured. According to the occurrence of pores in clay, the classification can be composed of four levels (particle, particle group, unit, and aggregate), and accordingly the pores are divided into five categories of interparticle pore, intergroup pore, intra-aggregate pore, inter-aggregate pore, and aggregate connection pore (shown in Fig. 4.10).

According to a large number of SEM images, the fourth layer of saturated mucky clay of Shanghai mainly has flocculated structure and honeycomb-flocculation structure. Clay mineral particles are mainly illite, and also some chlorite, montmorillonite, etc. And the structure element is mainly laminar (see Fig. 4.11) while

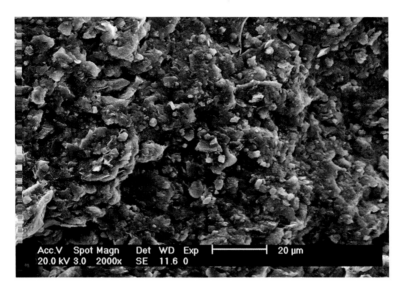

Fig. 4.11 SEM picture of undisturbed sample

the aggregate is flocculated, in flower and feather shapes (Tang et al. 2005). The connections of structure elements are mainly edge-to-face or edge-to-edge contact which forms the pore skeleton with high void ratio. This kind of electrostatic connection mode has certain strength, but the strength of connection is far less than the bonding links (such as Ca and Fe chemical bonding). So the connection is easy to be damaged with external force. However, the field test results show that at the depth of 8.5 and 11.5 m (close to the walls of the tunnel lining in the subway), the amplitude of soil earth pressure variation near outer wall of the tunnel lining induced by subway loading is 0.23 and 0.70 kPa which is a very tiny value. Although the connection strength of elements is not high, such small stress is not big enough to make severe damage. However, with the increase of cycle number, soil structure elements have a trend of consolidation. Two or more adjacent elements begin to approach and compress as the distance of the elements are close enough. The loosely bound water on the surface of structure elements is driven out so that strongly bound water on the surface of adjacent elements is connected directly. Because the strongly bound water attraction is quite huge (some scholars proposed that it can achieve 1,000 atm), when two structure elements come close without large external force, the compaction trend will increase quickly and become much firm (shown in Figs. 4.12 and 4.13). Pore water is difficult to flow from the space of these two elements, and this has significant influence on soil permeability. Therefore, these two elements are considered as a larger one. The structure element is mainly laminar while aggregate is flocculated. The vibration loads lead to more pores (Fig. 4.14), and this is why the quantity of larger pore increases rather than decreases. Because of the occurrence of pore structure, the total porosity has a rise trend.

4.2 Qualitative Analysis

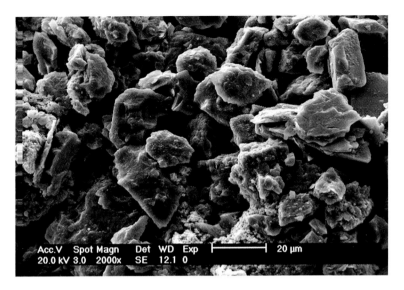

Fig. 4.12 SEM picture of NO. 830D1 sample after vibration

Fig. 4.13 SEM picture of NO. 826D2 sample after vibration

Based on the compaction trend of the structural elements, generally speaking, one surface vanishes because two lamellar structure elements have formed a larger element (two elements share one surface). This is the reason why the specific surface area have a downward trend under vibration.

However, at the tunnel bottom (about 13.5 m), the earth pressure amplitude is as large as 1.15 kPa caused by subway loading which is about 5 times and 1.6

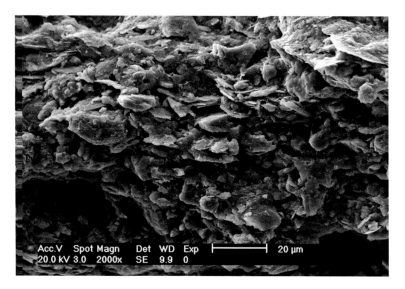

Fig. 4.14 SEM picture of NO. 828D3 sample after vibration

times of those near the outer walls of the tunnel lining (8.5 and 11.5 m). Because of the increase of cyclic stress, the unstable structure element which originally has a compaction trend starts slip movement (releasing the overlap surface). This is the reverse process of compaction. Thus, large pores have been filled by smaller structural elements or aggregates, and the component of large pore reduces. This leads to a tiny rebound of the total specific surface area and a slight decrease of uniform average pore diameter, total porosity, and the retention factor.

Based on the analysis above, it can be forecasted that during quite a long time of subway operation, the soil layer around the subway tunnel and tunnel bottom experience a compaction process. But because the pressure at the bottom of tunnel is large, the compaction trend has eased. The slip movement will be the main factor of deformation. Therefore, during quite a long time, deformation continues to increase, and the test results provide valuable materials for study on subway tunnel axis deformation of saturated soft clay in soft soil area and surface subsidence in tunnel area. Meanwhile, the microscopic test results also indicate that the packing model is not suitable for explaining the mechanism of deformation of saturated soft clay under low stress conditions.

In addition, combined with the dynamic triaxial test results in the previous chapter with SEM test results, some new progress of study on the microscopic mechanism of Shanghai saturated soft clay deformation and strength has been achieved.

Saturated soft clay is mainly porous loose honeycomb structure with high porosity ratio and moisture. When the subway cyclic loading have influence on soil, soil particles will suffer from certain inertia force which is actually a shear force caused by vibration. The surface of saturated soft clay particles has a bound water

4.2 Qualitative Analysis

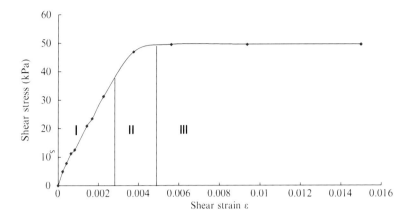

Fig. 4.15 Dynamic shear stress–strain curve

film forming certain cohesion. When the exerted shear force is very small (less than soil particle surface cohesion), soil particles do not move which means no excess pore pressure is produced, the soil is elastic in this process. That is to say, within a elastic stress area, the soil does not produce residual deformation when the stress ratio is less than the critical stress ratio and the cycle number N is very small (Zhou et al. 2000) (shown in Fig. 4.15). With the increase of loading, soil particles begin to slip and produce a certain deformation. When the shear stress continues to increase, more than a certain value (i.e., cyclic stress ratio is greater than the critical stress ratio), the connection strength and structure state will be damaged and the soil structure tend to be much looser. Meanwhile, some connections of particles are broken down, and the pressure (effective pressure) bearing by the contact point of soil particles at first is converted into pore water pressure resulting in the creation of excess pore water pressure. At this time if the load increase continues, residual deformation is produced because of the gradual destruction of soil strength (Hu et al. 2001).

Through the dynamic cycle triaxial test, the relationship between shear stress and shear strain is shown as Fig. 4.15.

Totally, the process of deformation can be divided into the following three stages:

Stage I: During this stage, because the shear stress is small, if the external force is relieved, the soil microstructure will be retained perfectly, and there is no obvious change of soil structure after vibration. The soil particle structure returns to its original state without residual deformation, and the performance of soil is elastic, so this stage is quasi-elastic stage.

Stage II: If the shear stress is larger than a certain value, the soil particles will slip as soil microstructure deformation appears. This is an energy accumulation process. When the shear stress energy accumulates to a certain extent and exceeds some soil particle connection energy, soil particles begin to slip with the damage of soil microstructure. Thus, soil residual deformation is formed and this stage reflects the soil elastic-plastic character; that is to say, this stage is elastic-plastic stage.

Stage III: If the shear stress continues to increase (stress ratio is greater than the critical stress ratio), the soil microstructure will completely be destroyed with decrease of the soil strength, and the strain tends to infinity. This stage is known as softening stage as soil is "softening." Due to the increase of the shear stress, the shear stress energy also has an upward trend. If the shear energy exceeds connection energy between major soil particles, the microstructure of soil is destroyed and connection force between soil particles disappears completely. With the vibration load, soil particles will move or rotate in horizontal or vertical direction until reaching a new stable equilibrium which is the "rearrangement" mentioned previously. According to the principle of energy, soil particles have certain orientation after rearrangement. Due to the effect of the vibrations, pore among soil particles is compacted and residual deformation forms.

With the growth in the loading cycle number, the soil structure strength tends to be fatigued and the soil inter-aggregate cohesive force decreases; then, the soil deformation generates. If the stress is not big enough and remains stable, the deformation tends to be a constant value which is consistent with the results of cyclic triaxial test.

According to the above analysis, during different stress stages, the deformation mechanism of soil microstructure is diverse. Also, this process can well confirm the cycle triaxial test results. It is clear that soil around the subway tunnel is in the first stage (quasi-elastic stage). If exceeding this stage, soil particle will slip and the soil microstructure will be partly damaged to induce some residual deformation. In addition, the tunnel axis will produce larger settlement which will affect the normal operation of the subway.

The buried depth of Shanghai Subway in saturated soft clay area is about 8–17 m, so the sample of this test is retrieved from the third layer of mucky silty clay and the fourth layer of mucky clay in Shanghai, which are exactly located at these depths. In terms of certain engineering project, monitoring of deformation surrounding subway tunnels is required within some critical value to guarantee the safety (the critical value is 0.0028–0.0033 derived from the test results). If the soil deformation is controlled in the first stage, the subway train will be kept in normal safe operation without residual deformation.

4.3 Quantitative Analysis

Under long-term cyclic loading, the soil deformation increases with the time lasting. The deformation is closely related to soil structure, i.e., soil deformation and displacement is mainly caused by the change of internal structure element variation (the soil deformation is the macro behavior of the accumulation of microslide in soil structure elements). All the structure elements remain in a dynamic balance during the process of deformation. The soil microstructure has unique characteristics in time and space domains. At present, most of the soil microstructure morphology researches are based on scanning electron microscopy (SEM) method. But SEM method itself has some defects. For example, it is difficult to accurately quantify SEM images without suitable image processing techniques. And the SEM images acquired are 2D so that it can be hardly used to determine the structure element variation or changes of similar soil in different engineering environment. SEM method can acquire relevant qualitative analysis for soil microstructure; mercury intrusion porosimetry (MIP) can provide access to quantified research on microstructure. Therefore, this part focuses on quantitative research of soil microstructure element variations caused by subway loading to explore the characteristics of pore size distribution, pore quantity, and other pore structure parameters before and after loading. Full explanation and validation of quantitative results of soft clay can be proposed combined with qualitative and visible analysis by SEM images.

4.3.1 Research Method

4.3.1.1 Mercury Intrusion Porosimetry (MIP)

This test apparatus is the American AUTOSCAN 60 automatic mercury intrusion porosimetry (MIP) shown in Fig. 4.16. The samples were prepared by freeze-dry dehydration for test and storage with plastic wrap as shown in Fig. 4.17. During the test, mercury was intruded into the sample tube under the vacuum condition. Low-pressure analysis (Fig. 4.18a) was conducted and then high-pressure analysis (Fig. 4.18b) was followed. The maximum high pressure can reach 60,000 psi. Usually during the mercury intrusion process, the range of typical apparent pore entrance diameters is about 2.25–4266.44 nm under pressure of 24–47,430 psi; and during the extrusion process of mercury, the range is 2.26–3677.97 nm under pressure of 29–47,161 psi.

When a liquid does not wet a porous solid, it will not enter the pores in the solid by capillary action. The non-wetting liquid (mercury in this test method) can be forced into the pores by the application of external pressure. Mercury is a liquid metal, and it has not only electrical conductivity but also liquid surface tension. And because of these features, in the process of mercury intrusion, with the increase of pressure, mercury is forced into pores in samples. Then the electrical signals are input into computer for data processing, and those related spectrums are simulated for calculation of porosity and specific surface area. The intruded size of pores is inversely proportional to the applied pressure. When a cylindrical pore model is assumed, the relationship between pressure and size is given as follows:

$$p\pi r^2 L = \gamma \cdot 2\pi L \cos\theta = p \cdot \Delta V \tag{4.1}$$

It can be transferred into Eq. (4.2) by transposition:

$$r = \frac{-2\gamma \cos\theta}{p} \quad \text{(Washburn's)} \tag{4.2}$$

Fig. 4.16 Automatic mercury analyzer AUTOSCAN 60

Fig. 4.17 Sample of mercury analyzer

where r is the pore diameter, θ is contact angle, p is pressure, γ is the surface tension of the mercury, L is the length of the pore, ΔV is intruded volume variation for mercury, and S is the pore surface area.

4.3 Quantitative Analysis

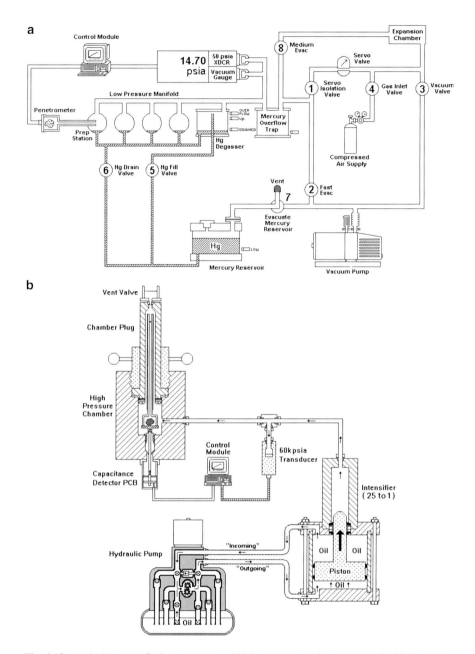

Fig. 4.18 Analysis system for low pressure and high pressure: (**a**) low pressure; (**b**) high pressure

The volume of the intruded pores is determined by measuring the volume of mercury forced into them at various pressures. The relationship between pore surface area and pressure needed to fill all the corresponding pore space with mercury is as follows:

$$S = \frac{p \cdot \Delta V}{\gamma \cos \theta} \tag{4.3}$$

If $\gamma/\cos \theta$ is constant ($\gamma = 0.473 - 0.485$ N/m, $\theta = 135° - 142°$ which is related to the purity of mercury; an average of 0.480 N/m and 140° can be taken respectively without much error), then

$$S = \frac{1}{\gamma \cos \theta} \int_0^V p dV \tag{4.4}$$

$$\text{Porosity} = 100 \left(\frac{V_a}{V_b} + \frac{V_a - V_b}{V_c - V_b} \right) \tag{4.5}$$

Where

V_a = the volume of mercury intrusion under any pressure
V_b = the volume of mercury intrusion in steady state
V_c = the volume of mercury intrusion with the largest pressure during measurement

According to Eq. (4.2), the pore diameter r is inversely proportional to the pressure p. Based on Eqs. (4.4) and (4.5), the sample's specific surface area and porosity are related to the mercury volume of intrusion. The specific surface area can be calculated by the pore diameter.

4.3.1.2 Fractal Research

With the development of fracture mechanics, porosity theory, and fractal theory, scholars have proposed that pore structure of some porous materials has significant fractal characteristics. For these materials, their pore structures, physical properties, and chemical properties can all be described by the index as pore fractal dimension. Soil actually has statistical self-similar fractal structure features (Mandelbrot 1982). With self-similarity statistical method to describe the distribution characteristics of complex soil pore quantitatively, the deformation properties and mechanical behavior can be revealed essentially (Wang et al. 2000). Our research group chose saturated soft clay near subway tunnel between Jing'an Temple Station and Jiangsu Road station of Line 2, Shanghai Metro. The study on fractal microstructure of saturated soft clay around the tunnel is based on the field monitoring and laboratory GDS test and results of MIP.

4.3.2 Fractal Parameters of MIP Results

The percentage of pore volume occupied by mercury decreases with the growth of mercury pressure from p to $p + \delta p$; in the tiny interval of the pore radius $(d, d + \delta d)$, the corresponding pore diameter reduces from r to $r + \delta r$. D_V is termed as the pore volume distribution rate relevant to the pore size; then

$$dV = -D_V(r)dr \tag{4.6}$$

where D_V is the pore volume variation induced by the change of pore size per unit. Equation 4.7 can be derived in conjunction with Eqs. (4.2) and (4.6):

$$D_V(r) = \frac{p}{r}\left(\frac{dV}{dP}\right) \tag{4.7}$$

Another equation used to analyze the pore distribution is the natural logarithmic pore size-based volume distribution function $D_V(\ln r)$, namely,

$$D_V(\log r) = \frac{dV}{d \log r} = rD_V(r) \tag{4.8}$$

Equations 4.2 and 4.8 form a complete relation of pore size distribution and mercury intrusion pressure, i.e.,

$$D_V(\log r) = p\frac{dV}{dp} = \frac{dV}{d \log p} \tag{4.9}$$

During the mercury intrusion process (Matthews et al. 1995), the pressure p used for mercury intruding is directly related to pore size distribution. Meanwhile, in the process of mercury extrusion, the pressure p is directly related to the pore size, i.e., when $dV/dr \leq 6$, mercury can flow out free; on the other hand, mercury remains in pore.

From a full mercury intrusion-extrusion curve, many parameters in specific size range can be attained, such as peak pore diameter, average pore size value, equivalent pore sizes, porosity, pore size distribution, specific surface area distribution, and pore quantity. The selection of parameters is based on the research purpose, and this can provide access to quantitative study of pore characteristics. As it is well known, pore structure of clay strongly affects the strength and permeability which is concerned in engineering area. So this paper studies on the characteristics of pore structure variation with influence of subway vibration loading from three aspects: total porosity, average pore size value, and the retention factor. Some definitions are as follows:

n_p (total porosity) refers to the ratio of total pore volume to the total sample volume, namely,

$$n_p = \frac{V_p}{V} \tag{4.10}$$

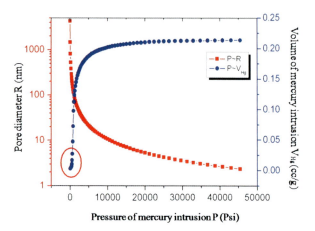

Fig. 4.19 Curve of pore diameter versus pressure of mercury intrusion

where V_p is pore volume, V is the total volume of soil sample, and n_p can be obtained from cumulative intrusion volume and the relevant weight.

For the pore mercury intrusion curve, the pore size range can be divided into n intervals. In the ith pore size range, V_i is the increment of mercury volume and r_i is the interval average radius. Mean distribution radius is given in Eq. (4.11):

$$\ln r_m = \frac{\sum_{1}^{n} V_i \ln r_i}{\sum_{1}^{n} V_i} \qquad (4.11)$$

R_f (retention factor) refers to the ratio of residual mercury volume in the sample and total mercury volume after test, namely,

$$R_f = \frac{V'_{Hg}}{V_{Hg}} \qquad (4.12)$$

where V'_{Hg} is the residual mercury volume and V_{Hg} is the total mercury volume.

In addition, in terms of general international pore size range definition, large pore is with radius over 50 nm, medium pore is with radius of 2–50 nm, and pore in microlevel is with radius less than 2 nm.

4.3.3 The Characteristics of Pore Variation in Undisturbed Soft Clay During the Process of Mercury Intrusion

In the process of mercury intrusion, the range of pressure is 25–47,430 psi, and the pore size detected is in the range of 2.25–4266.44 nm. As shown in Figs. 4.19,

4.3 Quantitative Analysis
133

Fig. 4.20 Curve of volume versus pressure of mercury intrusion

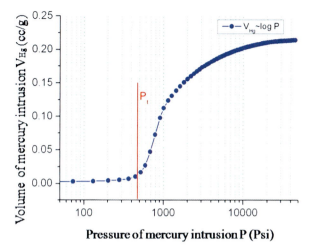

Fig. 4.21 Curve of specific surface area of accumulation versus pressure of mercury intrusion

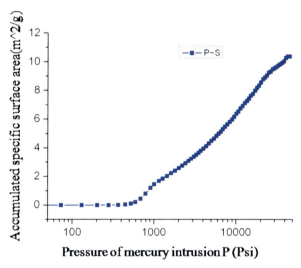

4.20, 4.21, 4.22, 4.23, 4.24, 4.25, and 4.26, in the mercury intrusion process, with the increase of intrusion pressure, the volume of mercury intruding into pores behaves an upward trend while the pore size has a diverse trend especially in the primary stage (Fig. 4.19). However, the growth rate of pore volume is very slow at the beginning and increases rapidly until a certain pressure reaching the peak (the pressure value is 366 psi, diameter is 291.42 nm). When the increasing rate of mercury intrusion volume reaches the peak, the corresponding pressure is 788 psi, the pore size is 135.36 nm, and peak value is 0.0254 cm^3/g. Then the speed tends to be stable. The reason is as follows: in the stage with low pressure, the mercury just intrudes large pore at first. After the pressure reaching a certain pressure (the threshold pressure value Pt), mercury begins to travel through narrow pore throat and the mercury volume change is apparent (shown in Fig. 4.20) which is just

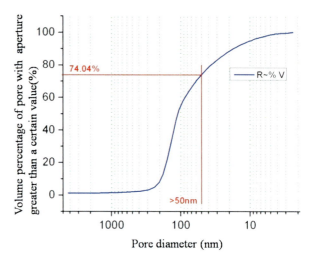

Fig. 4.22 Distribution of accumulation pores

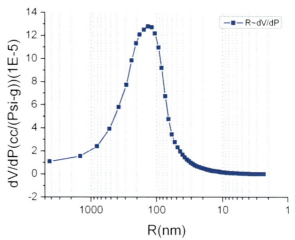

Fig. 4.23 Curve of differential pore volume distribution

as bottleneck effect. Since then, as the mercury pressure increases, the volume of mercury intrusion begins to increase rapidly and just with a tiny pressure, the volume can reach the peak. From then on, the P–V_{Hg} curve tends to be stable. It indicates that the quantity of intruding mercury is stable and the pore volume grows slowly.

At the same time, with the growth of mercury-intruding pressure, accumulated specific surface area continuously rises slowly at the beginning. But the growth rate tends to stable after approaching a threshold value Pt (Fig. 4.21). Even if the changing rate of mercury-intruding volume is different, pore-specific surface area filled by mercury does not have conspicuous change. It is mainly because of offset effects of pore quantities. That is to say, specific surface area is related with not only pore size but also number of pores. With the decrease of pore radius, the pore size decreases, but the total pore number increases. At last, the combining effect makes the specific surface area remain stable growth.

4.3 Quantitative Analysis

Fig. 4.24 Curve of $D_v(r)$ with pore distribution

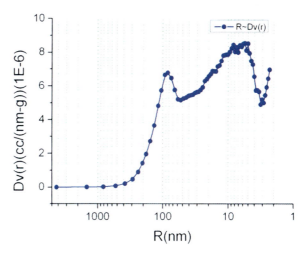

Fig. 4.25 Curve of $D_s(r)$ with pore distribution

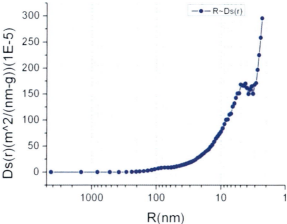

In addition, according to the mercury intrusion curve, the percentage of large pore volume of undisturbed soft clay can be obtained (Fig. 4.22). According to the curve, the pore volume percentage is 74.04 % when the pore size is greater than 50 nm. The others are macropores and micropore.

4.3.4 The Characteristics of Pore Variation in Undisturbed Soft Clay During the Process of Mercury Extrusion

In the process of mercury extrusion, the pressure is within the range of 29–47,161 psi, and the detected pore size is within the range of 2.26–3677.97 nm. The mercury intrusion curve is used to illustrate the variation characteristics of

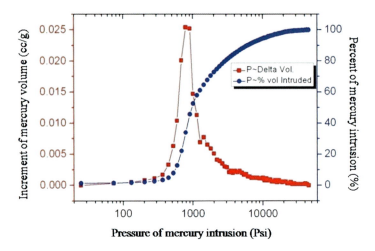

Fig. 4.26 Curve of volume percent versus pressure of mercury intrusion

mercury intrusion volume with the increase of pressure, while the mercury extrusion curve demonstrates the changing rule with the decrease of pressure. As shown in Fig. 4.27a, when reaching a certain internal pressure value, intrusion-extrusion mercury curve is not closed, which means some mercury has remained in the pore permanently. In fact, the path of mercury intrusion curve and mercury extrusion curve is not consistent. In terms of any given pressure value, the corresponding volume of mercury extrusion curve is higher than the mercury intrusion curve. Instead, given a certain volume value, the pressure value in mercury intrusion curve is higher as shown in Fig. 4.27b.

4.3.5 The Variation of Pore Structure Characteristic Parameters in Saturated Soft Clay Under Vibration Loading

4.3.5.1 Quantitative Analysis Results

With mercury intrusion results, Fig. 4.28 shows the pore size distribution curve of the gray mucky clay surrounding subway tunnel in fourth layer. Samples 830D1, 826D2, and 828D3 are all obtained from the GDS test. Specifically speaking, the depths of samples 830D1, 826D2, and 828D3 are 8.5 m, 11.5 m, 13.5 m, respectively. It can be seen that there is no conspicuous difference between pore size distribution curves of vibrated samples and undisturbed sample and the pore sizes are mainly between 100 and 300 nm. According to pore size scale mentioned above, obviously, the pore sizes in the samples are mainly large pores. In order to

4.3 Quantitative Analysis

Fig. 4.27 Curve of volume of intrusion (extrusion) versus pressure of mercury intrusion: (**a**) intrusion curve; (**b**) extrusion curve

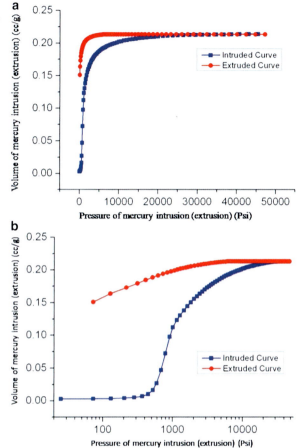

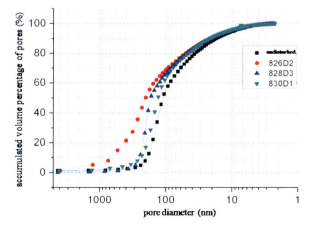

Fig. 4.28 Distribution of pore diameters

Table 4.1 Characteristic parameters of pore structure of undisturbed and vibrated samples

Sample number	n_p (cm³/g)	r_m (nm)	R_f	S (m²/g)	Pore distribution (%) <50 nm	>50 nm
Undisturbed	0.2138	76.66	0.7687	10.5134	26.0	74.0
830D1	0.2229	99.51	0.7981	9.6124	21.0	79.0
826D2	0.2352	153.40	0.8445	8.6740	19.9	80.1
828D3	0.2301	107.94	0.7565	9.4232	21.2	78.8

analyze the variation of microstructure parameters in samples, the statistical results of pore size distribution and other parameters by mercury intrusion porosimetry are shown in Table 4.1.

In terms of the pore size distribution of three groups of samples, statistical results show that the quantity of meso-pore (<50 nm) and micropore has decreased while the quantity of macropore has increased.

n_p of vibrated samples in three groups all marginally rise compared to undisturbed samples. With the increase of depth (or cyclic stress ratio, CSR), n_p increases initially and then tends to decrease; r_m of three groups of vibrated samples all augment compared to undisturbed samples dramatically. Along the depth, r_m behaves similarly to n_p. R_f of vibrated samples 830D1 and 826D2 slightly increases compared with undisturbed samples, while R_f of 828D3 has a little decrease. The surface areas of three vibrated samples all slightly reduced compared with the areas of undisturbed samples.

4.3.5.2 Variation and Mechanism Analysis of Characteristic Parameters of Pore Structure Before and After Subway Loading

Statistical results of the variation of microstructure parameters with mercury intrusion are shown in Table 4.2.

From the statistical results of pore size distribution, the quantity of pores within the range of smaller than 200 nm decreases in all vibrated samples compared to the undisturbed sample, while pore quantity in the range of 200–300 nm in samples 826D2 and 828D3 decreases dramatically. And pores larger than 200 nm increase a lot especially in the range of 200–300 nm; R_f in vibrated samples, 830D1 and 826D2, grow a little compared with undisturbed samples, while R_f of 828D3 has a little decrease.

According to a large number of SEM images, the saturated mucky clay of the forth layer in Shanghai mainly has flocculated structure and honeycomb-flocculation structure. The structure element is mainly laminar (Fig. 4.11) while the aggregate is flocculated, in flower and feather shape (shown in Fig. 4.8b). The connections of structure element are mainly edge to face and edge to edge which forms the pore open structure with high void ratio. This kind of electrostatic

4.3 Quantitative Analysis

Table 4.2 Characteristic parameters of pore structure of soil samples before and after vibration

Sample number	Total mercury volume V_{Hg} (cm³/g)	Residual volume V'_{Hg} (cm³/g)	Retention factor R_f	Total porosity n_p (cm³/g)	Specific surface area S (m²/g)	Mean distribution radius r_m (nm)
Undisturbed	0.2138	0.1645	0.7687	0.2138	10.5134	76.66
830D1	0.2429	0.1686	0.7981	0.2429	9.6124	99.51
826D2	0.2352	0.1836	0.8445	0.2352	8.6740	153.40
828D3	0.2301	0.1986	0.7565	0.2301	9.4232	107.94

	Pore size distribution (%)					
Sample number	<50 nm	50–100 nm	100–200 nm	200–300 nm	>300 nm	ε_a (%)
Undisturbed	26.9	19.8	45.2	5.6	3.4	–
830D1	21.0	15.5	44.1	12.4	7.0	0.11
826D2	19.9	11.1	15.6	23.8	29.6	0.17
828D3	21.2	12.8	23.4	28.4	4.2	4.50

connection has certain strength, but the strength of connection is far less than the chemical bonding (such as Ca and Fe cement). So the connection is easy to be damaged with the action of external force.

The occurrence of pore open structure increases the total porosity instead. At the same time, because of compaction effect on the structure element, for those edge-to-edge contact elements, one larger structure element emerged. It results in decline tendency of total specific surface area. However, at the tunnel bottom (around 13.5 m), the earth pressure amplitude is as large as 1.15 kPa caused by subway loading which is about 5 times and 1.6 times of those near the outer wall of the tunnel lining (8.5 and 11.5 m); corresponding cyclic stress ratio is 6 times and 2 times. Because of the increase of cyclic stress, the originally compacted structure begins to appear slip movement (releasing the overlap surface) with the cycle loading lasting, which mainly happens in the pore size of 50–200 nm. Thus, pores (50–200 nm) have been filled by smaller structural elements and aggregates. Because the large pores (>200 nm) barely change in this process, the decrease of the pores between 50 nm and 200 nm causes a slight rebound of the total specific surface area and a small decrease of uniform pore diameter as well as total porosity.

4.3.6 *Fractal Model and Fractal Dimension*

4.3.6.1 Fractal Model of Pore Characterization

The Menger Sponge Model

Considering dividing each side of a cube into b subside to form smaller cubes, random integers of $1 - b^3$ are set to number all these small cubes. Now some manipulation is applied to take away the cubes gradually as following rules. Assuming the probability of occurrence of pore during every manipulation is p ($0 < p < 1$), an integral number is taken from this formula $n = [(b^3 - p \times b^3)]$. If the number of a certain cube is larger than n, this cubes can be removed; otherwise, it can be kept. Repeating the above manipulation, the remaining solid cube dimension size is gradually decreasing till reaching the designed iterations. When finishing the above manipulation, the remaining hollow part represents the pore structure and the solid part represents the soil skeleton. The above manipulation rule is called the Menger random porous media model (Coppens and Froment 1995).

After *i* times iterations, even if the spatial distribution of different size pores is totally random, the number of solid cubes with size of $1/b^i$ follows a certain rule:

$$N\left(1/b^i\right) = \left(b^3 - p \times b^3\right)^i \qquad (4.13)$$

4.3 Quantitative Analysis

According to the law of scale invariant, solid cube number N $(1/b^i)$ can meet the following formula:

$$N(1/b^i) = k(1/b^i)^{-D_M} \tag{4.14}$$

where k is the number of original cubes and D_M is the fractal dimension based on Menger sponge model. Assuming $r_i = 1/b^i$, after i times iterations, the volume of solid skeleton V_s (m^3) will meet the following formula based on Eq. (4.14):

$$V_s \propto r_i^{3-D_M} \tag{4.15}$$

Because the pore volume $V_p = 1 - V_s$, if $i >> 1$ or $r_i << 1$, the relationship between pore volume V_p and pore size distribution r_i tends to be $V_p = f(r)$. Then Eq. (4.15) can be differential with respect to r as follows:

$$-\frac{dV_p}{dr} \propto r^{2-D_M} \tag{4.16}$$

From Eq. (4.16), the relationship between pore volume and pore size is essential for the calculation of fractal dimension of saturated soft clay. It can be obtained from Washburn's formula shown in Eq. (4.2). The fractal dimension value can be derived in natural logarithm relation between pore volume and pressure in conjunction with Eqs. (4.2) and (4.16), take the logarithm on both sides, and write an equation; the associated formula of fractal dimension is shown as follows:

$$\ln\left(\frac{dV_p}{dP}\right) = (D_M - 4)\ln P + C \tag{4.17}$$

where C is a constant.

The fractal dimension D_M of saturated soft clay can be derived from the slope value of $(D_M - 4)$ in line $\ln(dV_p/dP) \sim \ln P$, assuming the pore volume to be equal to the volume of intruding mercury.

Fractal Model Based on the Thermodynamic Relations

When measuring the relationship between pore volume and pore size of porous material, the external work on the mercury equals to the increase in surface energy of the mercury in the pores. And the pressure P (Pa) applied on the mercury and the volume of intrusion mercury V (m^3) meet the following equation:

$$\int_0^V PdV = -\int_0^S \sigma\cos\theta dS \tag{4.18}$$

Based on dimensional analysis, the pore fractal dimension can be established with respect to fractal scale of pore surface area S of porous materials by relating to pore size r, the volume of intrusion mercury V. For mercury intrusion operation, Eq. (4.18) can be written in a discrete form (Zhang and Li 1995):

$$\sum_{i=1}^{n} \overline{P}_i \Delta V_i = k r_n^2 \left(\frac{V_n^{1/3}}{r_n} \right)^{D_T} \tag{4.19}$$

where

$\overline{P}_i$ = the average pressure of ith mercury intrusion operation
ΔV_i = the volume of mercury during ith mercury intrusion operation
n = the interval number of pressure applied during ith mercury intrusion operation
r_n = pore radius corresponding the nth mercury intrusion operation
V_n = the accumulate volume of intrusion mercury with pressure interval $1-n$
D_T = pore fractal dimension based on the thermodynamic relations
k = coefficient constant

Defining

$$W_n = \sum_{i=1}^{n} \overline{P}_i \Delta V_i \tag{4.20}$$

$$Q_n = \frac{V_n^{1/3}}{r_n} \tag{4.21}$$

Equations (4.20) and (4.21) in conjunction with Eq. (4.19) form a complete linear line, and its logarithm form is shown as Eq. (4.22):

$$\ln \left(\frac{W_n}{r_n^2} \right) = D_T \ln Q_n + C_1 \tag{4.22}$$

where C_1 is a constant.

Equation 4.22 is associated with the applied pressure P and the volume of mercury in the process of mercury intrusion. Setting $\ln Q_n$ to be the horizontal ordinate and $\ln(W_n/r_n^2)$ to be vertical ordinate, the linear slope value should be the fractal dimension D_T of saturated soft clay.

4.3.6.2 Calculation of Fractal Dimension

Processing the MIP data according to Eq. (4.17), firstly the fitting results of fractal dimension of Menger sponge random model are shown in Figs. 4.29, 4.30, 4.31, and 4.32. From the figures, it can be seen distinctly that the relation of $\ln(dV_p/dP) \sim \ln P$

4.3 Quantitative Analysis

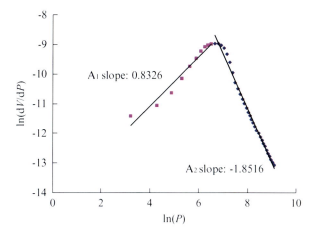

Fig. 4.29 Fractal dimension of undisturbed sample calculated through Menger model

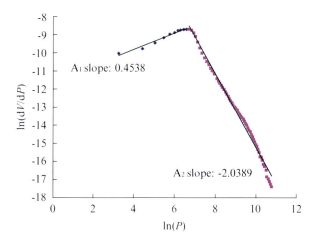

Fig. 4.30 Fractal dimension of undisturbed sample 830D1 calculated through Menger model

is not linear; and it is difficult to get rational fractal dimension value by simple linear fitting. There is obvious inflection point around the pore size of 150 nm. Hence, the experimental data are all fitted in piecewise dividing by the demarcation point of 150 nm. The D_M value derived from the fitting results of slope value is presented in Table 4.3.

The logical range of pore structure fractal dimension D_M of Menger sponge random model is 2–3. Larger fractal dimension indicates that the pore surface is far from smooth surface and the pore structure is more complex (Liu et al. 2004). According to Table 4.3, when the pore radius is >150 nm, the range of D_M is about 3.724–4.833; when the pore radius is <150 nm, the range of D_M is about 1.961–2.236, and $\ln(dV_p/dP) \sim \ln P$ has no liner relationship. Actually, the values in

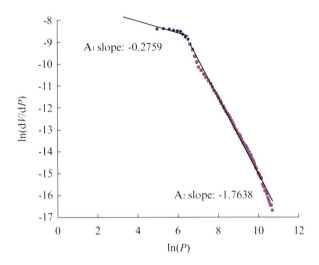

Fig. 4.31 Fractal dimension of undisturbed sample 826D2 calculated through Menger model

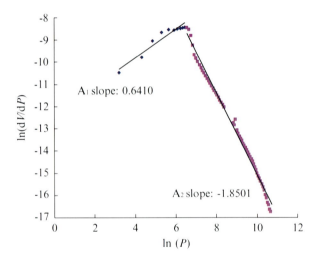

Fig. 4.32 Fractal dimension of undisturbed sample 828D3 calculated through Menger model

Table 4.3 are not reasonable in fractal dimensional analysis. It indicates that the Menger sponge random model is not appropriate for the pore fractal dimension analysis of saturated soft clay in the whole pore size range. So in the following part, the fractal dimension is calculated based on thermodynamic relations.

Taking $\ln Q_n$ to be the horizontal ordinate and $\ln(W_n/r_n^2)$ to be the vertical ordinate, the slope of the linear line is the fractal dimension D_T of saturated soft clay shown in Figs. 4.33, 4.34, 4.35, and 4.36.

4.3 Quantitative Analysis

Table 4.3 Fractal dimension calculated through Menger model

Sample number	Segment	r (nm)	Slope	D_M
Undisturbed	A1	>150	0.833	4.833
	A2	<150	−1.852	2.148
830D1	A1	>150	0.454	4.454
	A2	<150	−2.039	1.961
826D2	A1	>150	−0.276	3.724
	A2	<150	−1.764	2.236
828D3	A1	>150	0.640	4.640
	A2	<150	−1.850	2.150

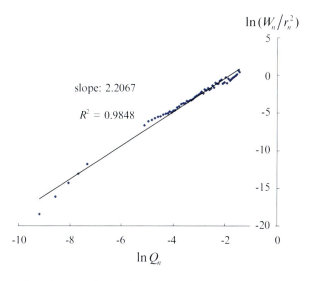

Fig. 4.33 Fractal dimension of undisturbed sample calculated through the thermodynamics model

According to Figs. 4.33, 4.34, 4.35, and 4.36, the correlation coefficients of mercury intrusion porosimetry data of saturated soft clay are above 0.98 which shows significant linear relationship. The fractal dimension is shown in Table 4.4. The fractal dimensions of undisturbed sample and vibrated samples are all in the range of 2.013–2.267. The relationship between fractal dimension D_T of thermodynamic relations model and cyclic stress ratio CSR is depicted in Fig. 4.37.

Fractal dimension is applied to reflect the complexity degree of pore space distribution for a certain material. That is to say, more complex pore space distribution leads to larger fractal dimension. According to Fig. 4.37, fractal dimension has a downward trend at first and then tend to increase with the cyclic stress ratio CSR. This phenomenon is caused by the following reason: when the CSR is very low,

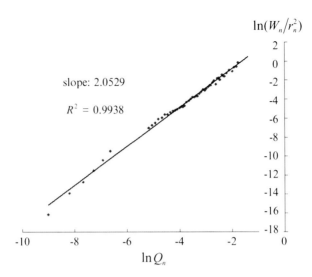

Fig. 4.34 Fractal dimension of sample 830D1 calculated through the thermodynamics model

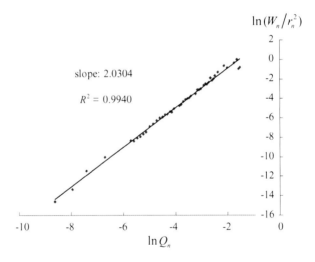

Fig. 4.35 Fractal dimension of sample 826D2 calculated through the thermodynamics model

the soil particle arrangement tends to be more oriented and dense, resulting in the decrease of fractal dimension. It indicates that the pores are more oriented and regular under low stress condition. On the other hand, with the increase of CSR, soil particles begin to slip and break down. Reflecting on the increase of fractal dimension, it shows that the porosity distribution becomes more complicated.

4.4 Correlation Analysis Between Microstructure and Macroscopic Deformation...

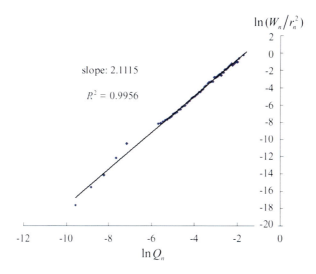

Fig. 4.36 Fractal dimension of sample 828D3 calculated through the thermodynamics model

Table 4.4 Fractal dimension calculated through the model of thermodynamics correlation

Sample number	Undisturbed	830D1	826D2	828D3
D_T	2.207	2.053	2.030	2.112
R^2	0.9848	0.9938	0.9940	0.9956

4.4 Correlation Analysis Between Microstructure and Macroscopic Deformation Properties of Saturated Soft Clay Under Subway Loading

4.4.1 Correlation Analysis of Pore Microstructure Parameter and Macroscopic Force

According to field test results, the variations of pore water pressure and earth pressure are relevant to the soil depth under subway loading; in essence, different depth means different stress condition. Thus, comprehensive analysis of correlation between pore microstructure parameters and cyclic stress ratio facilitates to the exploration of the variation mechanism of pore microstructure parameters (Tang et al. 2003). Figures 4.38, 4.39, 4.40, and 4.41 show the relationship between CSR and various pore microstructure parameters, respectively.

From Figs. 4.38, 4.39, and 4.40, it can be easily seen that when CSR is very low, the mean distribution radius r_m and total porosity n_p increase with the growth of CSR. But after the peak of n_p and r_m appears, both of them begin to decrease.

Fig. 4.37 Relationship between fractal dimension D_T and CSR

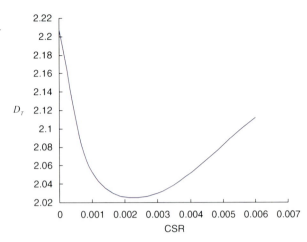

Fig. 4.38 Relationship between cyclic stress ratio and mean distribution radius

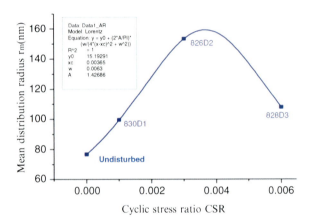

Meantime, the peak CSR is in the range of 0.003–0.004, and the trend is just like Lorentz curve. The variation-specific surface area behaves oppositely as above. The specific surface area decreases with CSR at first and then shows an upward trend after it reaches a valley value. Meanwhile, the inflection point of CSR is about 0.003–0.004 shown in Fig. 4.40.

Figure 4.41 shows that the curve of retention factor acts to be a quasi-symmetrical "saddle" shape. The inflection point of CSR also appears around 0.003–0.004. The rates of falling and rising in retention factor are almost equal in amount. The variation of pore microstructure parameters shows that a critical value of subway loading exists, and it plays an important role in the change of soil microstructure. The variation properties suggest that the deformation increases slowly when the load is less than critical value while the deformation increases sharply when the load is larger than the critical value.

4.4 Correlation Analysis Between Microstructure and Macroscopic Deformation...

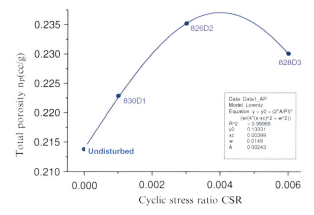

Fig. 4.39 Relationship between cyclic stress ratio and retention factor total porosity

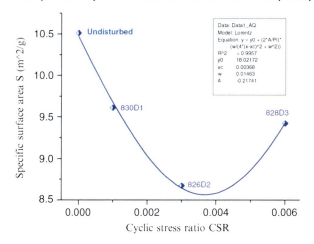

Fig. 4.40 Relationship between cyclic stress ratio and specific surface area

4.4.2 Correlation Analysis of Pore Microstructure Parameters and Macroscopic Deformation

Figures 4.42, 4.43, 4.44, and 4.45 present the relationship between pore microstructure parameters and macroscopic deformation. From the figures, the pore size and retention factor rise with increasing axial deformation. When the CSR is larger than the critical value (0.003–0.004), the axial strain increases rapidly while the mean distribution radius and retention factor have a downward trend (Zhao 2006). Specific surface area decreases with increasing axial strain and then tends to increase after over the critical CSR value. Total porosity has an upward trend after deformation arises. But due to the compaction effect, the total porosity begins to decrease gradually.

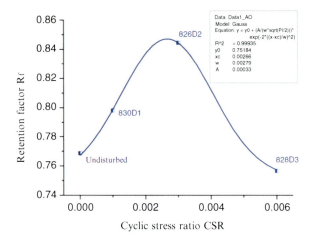

Fig. 4.41 Relationship of cyclic stress ratio and retention factor

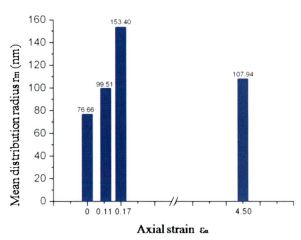

Fig. 4.42 Relationship of mean distribution radius and axial strain

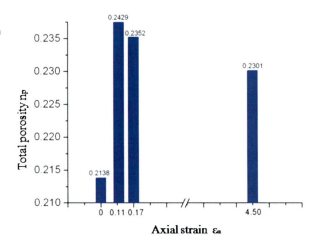

Fig. 4.43 Relationship of total porosity and axial strain

4.5 Chapter Summary

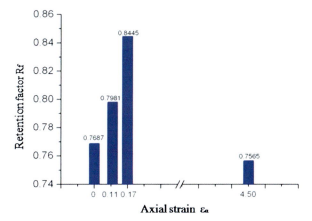

Fig. 4.44 Relationship of retention factor and axial strain

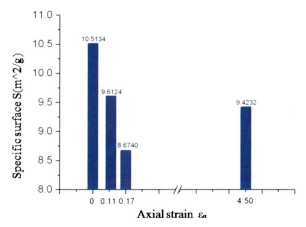

Fig. 4.45 Relationship of specific surface and axial strain

4.5 Chapter Summary

As a matter of fact, the macroscopic deformation of saturated soft clay under subway loading is a reconstruction process of microstructure. All the structure elements including particles and pores change under subway loading. According to the above analysis, the rearrangement state of microstructure under subway loading is related to the vibration duration and stress level. Based on SEM and MIP tests, fractal theory analysis, and correlation analysis of pore microstructure parameters and macroscopic characteristics, some conclusions can be drawn:

1. The saturated mucky clay of the fourth layer in Shanghai mainly has flocculated structure and honeycomb-flocculation structure. Clay mineral particles are major illite, less chlorite, and tiny montmorillonite. The shape of structure elements are mainly laminar (see Fig. 4.11), while the aggregates are flocculated in flower or feather shape. The connections of structure elements are mainly edge to face and edge to edge which forms an open pore structure with high void ratio.

2. The saturated mucky clay of Shanghai has primary pore structure of large inter-aggregate pores (radiuses are over 50 nm); it has pore throat effect in the process of mercury intrusion.
3. The variation features of pore structure parameters with different depth can be indicated by the cyclic stress ratio (CSR). With the increase of depth, the CSR grows while the specific surface area decreases at first and slightly rebounds later. The mean distribution radius, retention factor, and porosity have the opposite trend.
4. The microscopic test results indicate that packing model is not suitable for explaining the mechanism of deformation in saturated soft clay under low stress conditions.
5. Based on the fractal theory, the study on clay pore distribution shows that Menger sponge random model is not appropriate to describe the fractal characteristic of mucky clay in the whole size range. But the model based on thermodynamic relations is suitable. The relationship between fractal dimension calculated by thermodynamic relations model and CSR was demonstrated as well as explanations of the variation features.
6. The correlation analysis of pore microstructure parameters and macroscopic deformation (macroscopic force) shows that a critical value of subway loading plays an important role in the microstructure variations of mucky clay near subway tunnel.
7. During a long time since the first operation of subway, soil structure elements of soil layers near subway tunnel and at the bottom of track tend to be compacted. However, the pressure applied on soil at the bottom of subway tunnel is quite large as the trend is alleviated. The rearrangement of soil elements is the main reason of deformation. Therefore, the accumulation of deformation will continue for a long time.

References

Bai B, Zhou J (2001) The applications and advances of SEM in geotechnical engineering. J Chin Electron Microsc Soc 20(2):154–160

Brewer R (1964) Fabric and mineral analysis of soils. Wiley, New York

Casagrande A (1932) The structure of clay and its importance in foundation engineering. Reprint in: Contribution to soil mechanics, 1925–1940, Boston Society of Civil Engineers, pp 72–125

Chen (1957) Structure mechanics of clays. Academic Sinica Press, Beijing, pp 83–97

Chen GD (2001) International achievements of study on frozen soil mechanics and engineering. Adv Earth Sci 6(3):293–299

Chen JO, Ye B, Guo SJ (2000) Preliminary study on the microstructure and engineering property of clay in Pearl River Delta. Chin J Rock Mech Eng 19(5):674–678

Coppens MO, Froment GF (1995) Diffusion and reaction in a fractal catalyst pore–II diffusion and first order reaction. Chem Eng Sci 50(6):1027–1039

Deng J, Wang LM, Zhang ZZ et al (2005) Analysis on microstructure of the recently cumulated collapsible loess in Yongdeng County, Gansu Province. Northwest Seismol J 27(3):267–271

Du ZD (1990) Observation of loess microstructure using SEM. Radiat Prot Bull 3:22–25

References

Goldschmidt VM (1926) Naturwissenschaften 14:477

Gong SL (2002) The microscopic characteristics of Shanghai soft clay and its effect on soil mass deformation and land subsidence. J Eng Geol 10(4):378–384

Hu RL (1995) Study on microstructure quantitative model and engineering geological characteristics of clay. Geological Publishing House, Beijing

Hu XD (2002) Determination of load on frozen soil wall in unloaded state. J Tongji Univ Nat Sci Ed 30(1):6–10

Hu RL, Yeung MR, Lee CF et al (2001) Mechanical behavior and microstructural variation of loess under dynamic compaction. Eng Geol 59(3–4):203–217

Lambe TW (1953) The structure of inorganic soil. ASCE Proc 79:1–49, Separate 315

Li Q, Bao SS (2003) The computation of the porosity of saturated clay based on the gray scale of soft microstructure graph. J Chongqing Teach Coll Nat Sci Ed 20(1):30–33

Liu YZ, Chen SQ, Sun H (2004) Characterizing pores in freeze–dried materials by fractal models and fractal dimensions. Trans Chin Soc Agric Eng 20(6):41–45

Mandelbrot BB (1982) Fractal geometry of nature. Freeman, San Francisco

Matthews GP, Ridgway CJ, Spearing MC (1995) Void space modeling of mercury intrusion hysteresis in sandstone, paper coating, and other porous media. J Colloid Interface Sci 171:8–27

Sergeyev YM, Grabowska-Olszewska B, Osipov V, Sokolov V et al (1980) The classification of microstructures of clay soils. J Microsc 120(12):237–260

Shan HX, Liu YY, Jia YG, Xu GH (2004) Case study of microstructure transform of silt due to wave action on subaqueous slope of Yellow R., China. Chin J Geotech Eng 26(5):654–658

Shi B (1997) A simple quantitative analysis method for microstructure of clayey soil. Hydrogeol Eng 20(6):7–10

Shi B, Jiang HT (2001) Research on the analysis techniques for clayey soil microstructure. Chin J Rock Mech Eng 20(6):864–870

Silverman EN, Bates TF (1960) X-ray diffraction study of orientation in Chattanooga Shale. Am Miner 45:60–68

Tang YQ, Huang Y, Ye WM, Wang YL (2003) Critical dynamic stress ratio and dynamic strain analysis of soils around the tunnel under subway train loading. Chin J Rock Mech Eng 22(9):1566–1570

Tang YQ, Zhang X, Zhou NQ, Huang Y (2005) Microscopic study of saturated soft clay's behavior under cyclic loading. J Tongji Univ 33(5):626–930

Wang Q, Wang JP (2000) A study on fractal of porosity in the soils. Chin J Geotech Eng 22(4):496–498

Wang Q, Wang FY, Xiao SF (2001) A quantitative study of the microstructure characteristics of soil and its application to the engineering. J Chengdu Univ Technol 28(2):148–153

Wang BJ, Shi B, Liu ZL et al (2004) Study on microstructure fractal characteristics of clay based on GIS. Chin J Geotech Eng 26(2):244–247

Wang GX, Huang HW, Xiao SH (2005) Experimental study on micro-structural characteristics of soft soil. J Hydraul Eng 36(2):190–196

Wang BJ et al (2008) 3D visualization and porosity computation of clay soil SEM image by GIS. Rock Soil Mech 29(1):251–255

Wu YX (1991) Quantitative analysis of engineering clay microstructure. Acta Geosci Sin 23:143–153

Xiao SF, Akilov K (1991) The fabric and strength creep character of argillaceous intercalated bed. Science and Technology Jiling Press, Changcun

Xie HP (1992) Fractal geometry and its application to rock and soil materials. Chin J Geotech Eng 14(1):14–24

Zhang MY (1986) Study on soil microstructure using SEM. Rock Soil Mech 7(1):53–58

Zhang BQ, Li SF (1995) Determination of surface fractal dimension for porous media by mercury porosimetry. Ind Eng Chem Res 34:1383–1386

Zhao SK (2006) The study on the micro-structure distortion mechanics of soft clay under the subway loading. Department of Geotechnical Engineering College of Civil Engineering, Tongji University, Shanghai

Zhou CY, Mu CM (2005) Relationship between micro-structural characters of fracture surface and strength of soft clay. Chin J Geotech Eng 27(10):1136–1141

Zhou J, Gong XN, Li JQ (2000) Experimental study of saturated soft clay under cyclic loading. Ind Constr 30(11):43–47

Chapter 5
Finite Element Modeling

5.1 Introduction

The underground space is characterized as infinite and complicated. In the past decades, many scholars have done a lot of research on the ground vibration. They obtained analytical solutions to some idealized models and applied them into the qualitative and quantitative analysis of ground vibration problems. However, the physical and mechanical properties of soil are complex and inhomogeneous in the real infinite underground space; these simplified analytical models can hardly simulate the dynamic behaviors of the underground space accurately. Therefore, researchers and engineers started to focus their effort on the development of the numerical computation, especially on the finite element method (FEM), which has attained some satisfying preliminary achievements. In the numerical study of the underground space vibration induced by the subway trains, 2D (Wang et al. 1999) and 2.5D (Xie and Sun 2003) models are applied more frequently. Due to the complexity of the real engineering and limitation of the computation speed, 3D numerical model is less adopted. Andersen and Jones (2006) developed 2D and 3D FEM models with the boundary element method (BEM). They compared the solutions of these two models on frequency domain. The results demonstrated the 2D FEM model can only be used in the qualitative analysis of the underground vibration induced by subway trains.

So far, FEM used in the comprehensive analysis of underground vibration induced by transient excitation is rarely documented from practical view. Currently, it primarily establishes the physical model and finite element analytical model on the basis of soil natural properties, to simulate and solve the transient wave equation. Or FEM software (such as Super SAP, ANSYS, NASTRAN, and ABAQUS) is applied to analyze the problems of underground vibration, which can help us to obtain the stress distribution and other meaningful conclusions. In general, FEM provides an access to the comprehensive and accurate study on the underground vibration by the practical modeling method and computational methodology.

Some scholars (Balendra et al. 1991; Pan et al. 1995; Xia et al. 2005; Yan et al. 2006; Zhang and Pan 1993) conducted field dynamic tests and developed numerical models to study the vibration of subway system. However, due to the discrepancies of geological environment and subway system, these research findings cannot be applied in Shanghai area directly. In this chapter, we developed a finite element model for analyzing the dynamic response of the soil around the tunnel under the dynamic load induced by the subway. Through the finite element analysis, we obtained and visualized the instantaneous stress and the strain distribution in soil.

5.2 Finite Element Analysis Software

5.2.1 Numerical Method

In the practical engineering, we constantly find it impossible to obtain exact solutions to many problems. This can be attributed to the complexity of governing differential equations or the difficulties that arise from dealing with the boundary and initial conditions. To deal with these problems, we need to apply numerical methods. The first step of any numerical procedure is to divide the medium into a number of small elements and nodes, which is called discretization. There are two common numerical methods: finite differential methods (FDM) and finite element methods (FEM). In finite difference methods, the differential equation is written for each node, and the derivatives are replaced by difference equations. This approach results in a set of simultaneous linear equations. Although finite difference methods are easy to understand and employ in simple problems, they are difficult to apply to problems with complex geometrics, complex boundary conditions, and non-isotropic material properties. On the contrast to finite differential method, the finite element method uses integral formulations to create a system of algebraic equations. Then, a continuous function is assumed to represent the approximate solution for each element. The complete solution is generated by assembling the individual solutions.

5.2.2 The Finite Element Method and ANSYS

The finite element method can be used to obtain solutions to a large range of engineering problems involving stress analysis, heat transfer, electromagnetism, and fluid flow. It can effectively analyze the steady, transient, linear, or nonlinear problems of different fields or coupled fields. Undoubtedly, it is a multifunctional and widely compatible method in the engineering study and design. In different kinds of finite element analysis software, ANSYS is one of the most powerful and flexible software which can solve a variety of problems. ANSYS was released firstly in 1971 and has been a leading finite element analysis (FEA) program for almost 40 years. The current version of ANSYS has a very user-friendly interface (Saeed 2008).

In FEA software, the preprocessor provides two solid modeling approaches (top-down and bottom-up) and the functions of dragging, extending, rotating, moving, and copying the metafiles of solid model. Meanwhile, it provides four different methods of meshing (extending meshing, mapped meshing, free meshing, and adaptive meshing) to make this procedure more convenient. The solution processor has the commands that allow you to apply boundary conditions and loads. Once all the information is made available to the processor, it solves for the nodal solutions. The postprocessor includes general postprocessor and time-history postprocessor. The former one is usually used for analyzing static and steady-state problems; it can check the computational results of the entire model under a given load or time step or sub-step. The latter one can check the computational results of a given node which are dependent on time, frequency, or other factors, which means it is usually used for analyzing transient or nonlinear problems. Through the postprocessors, you can review the results of your analysis through graphical displays and tabular listings.

The dynamic structural analysis in FEA software is primarily developed on the general equation of motion in finite element system. It provides explicit and implicit integration method, which can be applied in the study of instantaneous dynamic response, modal analysis, harmonic response, and random vibration. The built-in nonlinear control system can make the most problems converging without extra operations from the users.

Moreover, the finite element analysis software can simulate the nonlinear properties of different materials, including plasticity, nonlinear elasticity, hyper-elasticity, viscoelasticity, and visco-plasticity. The Drucker-Prager (DP) model in finite element analysis software can better simulate granular material, such as rock or soil (Wang and Shao 1997).

5.3 Theoretical Analysis

Generally speaking, soil consists of solid particles, liquid, and gas; the solid particles constitute the soil skeleton and the liquid/gas constitutes the pore fluid. The properties of soil are controlled by the interaction of soil skeleton and pore fluid. For the soil, the train loading can lead to the increase, diffusion, and dissipation of pore water pressure; meanwhile, it can also cause the generation and development of residual strain. Due to the increase of pore water pressure and residual strain, the effective strength of soil decreases, which may cause massive deformation of soil and threaten the safety of the train operation.

For revealing the dynamic characteristics of soil under the train loading, Biot's consolidation theory was applied in the comprehensive analysis of the deformation and consolidation. We also developed a coupled dynamic model of train, rail, tunnel lining, and surrounding soil for studying the dynamic response of this system. More specifically, we introduce the pore water pressure and residual strain variables caused by the traffic loading into the Biot's equations. Through the FEM, the development of pore water pressure and residual strain can be obtained. In this process, the dynamic analysis is comprehensively considered with the dynamic seepage and skeleton deformation.

5.3.1 General Equations of the Dynamic Analysis

5.3.1.1 Model for the Dynamic Pore Water Pressure

For applying the effective stress in dynamic analysis, the model of the dynamic pore water pressure is necessary. The empirical formula for analyzing the dynamic pore water pressure is given as (Xu and Shen 1981)

$$\frac{u}{\sigma'_0} = \frac{\alpha}{\pi} \arcsin \left(\frac{N}{N_f} \right)^{\frac{1}{2\theta}} (1 - m\alpha) \tag{5.1}$$

where

σ'_0 = initial effective consolidation pressure
u = dynamic pore water pressure
N = the number of vibrations
N_f = the critical number of vibrations inducing the liquefaction
α, m, θ = empirical constants

5.3.1.2 Dynamic Constitutive Relation

The parameters which are required to develop Hardin-Drnevich model are just several, which make this model more simple and practical. Due to this, the material kernel curve of Hardin-Drnevich model is applied to define the constitutive relation:
  Shear modulus:

$$G = \frac{G_0}{(1 + \gamma_d/\gamma_r)} \tag{5.2}$$

$$G = \frac{E}{2(1 + \mu)} \tag{5.3}$$

  Equivalent damping ratio:

$$D_e = D_{e\ max} (1 - G/G_0) \tag{5.4}$$

5.3 Theoretical Analysis

where

G_0 = initial shear modulus
γ_d = dynamic shear strain
$\gamma_r = \tau_f/G_0$ = reference strain
τ_f = shear strength of soil
E, μ = elastic modulus and Poisson's ratio
$D_{e\ max}$ = the maximum equivalent damping ratio

Under the assumption of plane strain, the relation between strain and stress can be expressed as

$$\sigma_x = (\lambda + 2G)\varepsilon_x + \lambda\varepsilon_y$$
$$\sigma_y = \lambda\varepsilon_x + (\lambda + 2G)\varepsilon_y$$
$$\sigma_z = \mu(\sigma_x + \sigma_y)$$
$$\tau_{xy} = G\gamma_{xy} \tag{5.5}$$

where

$\sigma_x, \sigma_y, \sigma_z$ = the effective stress in the direction of x, y, z
$\varepsilon_x, \varepsilon_y, \gamma_{xy}$ = the normal and shear strain of soil skeleton in the direction of x and y

5.3.1.3 Geometric Equation

Under the assumption of small strain, the strain-stress relation can be expressed as

$$\varepsilon_x = -\frac{\partial u}{\partial x}$$
$$\varepsilon_y = -\frac{\partial v}{\partial y}$$
$$\gamma_{xy} = -\left(\frac{\partial u}{\partial y} + \frac{\partial v}{\partial x}\right) \tag{5.6}$$

where u, v = the displacement of soil skeleton in the direction of x and y.

5.3.1.4 Equilibrium Equation

According to Biot's consolidation theory, the skeleton of saturated soil is assumed as a linear elastic material. The equilibrium equation in the process of consolidation can be obtained through equilibrium analysis of a differentiation element which is taken from the soil. If the soil is under vibration, the pore water pressure will gradually increase; meanwhile, the pore water pressure is diffusing and dissipating. The equilibrium equation can be expressed as

$$\frac{\partial (\sigma_x + p + u_g)}{\partial x} + \frac{\partial \tau_{xy}}{\partial y} = \rho \ddot{u}$$

$$\frac{\partial \tau_{xy}}{\partial x} + \frac{\partial (\sigma_y + p + u_g)}{\partial y} + \rho g = \rho \ddot{v} \quad (5.7)$$

where

p = residual pore water pressure
u_g = dynamic pore water pressure
ρ = the density of saturated soil
g = gravitational acceleration
$\ddot{u}, \ddot{v}$ = the vibration acceleration of soil in the direction of x and y

Substituting formula (5.6) into formula (5.5) and then substituting the resultant new formula into formula (5.7), the stress equilibrium equations can be expressed by displacement u and v:

$$(\lambda + 2G) \frac{\partial^2 u}{\partial x^2} + G \frac{\partial^2 u}{\partial y^2} + (\lambda + G) \frac{\partial^2 v}{\partial x \partial y} - \frac{\partial p}{\partial x} - \frac{\partial u_g}{\partial x} - \rho \ddot{u} = 0$$

$$(\lambda + 2G) \frac{\partial^2 v}{\partial y^2} + G \frac{\partial^2 v}{\partial x^2} + (\lambda + G) \frac{\partial^2 u}{\partial x \partial y} - \frac{\partial p}{\partial y} - \frac{\partial u_g}{\partial y} - \rho \ddot{v} + \rho g = 0 \quad (5.8)$$

5.3.1.5 Continuity Equation of Seepage

According to Darcy's law, the flow rate of fluid while it passes the surface of a soil element can be expressed as

$$\begin{aligned} -\overline{k_x} \frac{\partial p}{\partial x} &= q_x \\ -\overline{k_y} \frac{\partial p}{\partial y} &= q_y \end{aligned} \quad (5.9)$$

where

$\overline{k_x} = k_x / \gamma_w$
$\overline{k_y} = k_y / \gamma_w$
k_x, k_y = the permeability coefficients of soil in the direction of x and y
γ_w = the unit weight of water
q_x and q_y are the flow rate of pore water in the direction of x and y

5.3 Theoretical Analysis

According to the continuity of saturated soil, the compressed volume of a soil element in unit time equals the volume of water which flows through the surface of this element, namely,

$$\frac{\partial q_x}{\partial x} + \frac{\partial q_y}{\partial y} = \frac{\partial}{\partial t}(\varepsilon_x + \varepsilon_y) \tag{5.10}$$

Substituting formula (5.6) and formula (5.8) into formula (5.9), the continuity condition can be expressed by displacement u and v:

$$-\frac{\partial}{\partial t}\left(\frac{\partial u}{\partial x} + \frac{\partial v}{\partial y}\right) + \overline{k}_x \frac{\partial^2 p}{\partial x^2} + \overline{k}_y \frac{\partial^2 p}{\partial y^2} = 0 \tag{5.11}$$

Combining formula (5.8) and formula (5.11), we obtained the equation set which can reflect the generation, diffusion, and dissipation of pore water pressure. The equation set refers three unknowns (u, v, p), which can be solved with some initial conditions and boundary conditions. According to the results, the displacement and residual pore water pressure of an element at any moment can be obtained.

5.3.2 Dynamic Finite Element Analysis

5.3.2.1 Numerical Analysis of Elements

According to the equilibrium equation (5.8), continuity equation (5.11), and dynamic pore water pressure equation (5.1), the unknowns (u, v, p) can be derived out by FEM. Firstly, unknowns can be substituted by the following functions:

$$u = \sum_{i=1}^{n} N_i(x, y) u_i(t)$$

$$v = \sum_{i=1}^{n} N_i(x, y) v_i(t)$$

$$p = \sum_{i=1}^{n} N_i(x, y) p_i(t)$$

$$\ddot{u}(t) = \sum_{i=1}^{n} N_i(x, y) \ddot{u}_i(t)$$

$$\ddot{v}(t) = \sum_{i=1}^{n} N_i(x, y) \ddot{v}_i(t) \tag{5.12}$$

$$N_i = \frac{1}{4(1 + \xi_i \xi)(1 + \eta_i \eta)} \tag{5.13}$$

where

N_i = the shape function in xy plane, and its value equals 1 at i node, while equals 0 at other nodes;
n = the amount of element nodes; and u_i, v_i,
p_i = the displacement and pore water pressure of the ith node at different times in any quadrilateral element.

According to the principle of virtual work, the partial differential equations $Lu = f$ are equivalent to

$$(Lu, f, A) = 0 \tag{5.14}$$

where

L = functional operation of u
A = an arbitrary virtual strain

For an arbitrary element e, given

$$A = \sum_{i=1}^{4} N_i A_i \tag{5.15}$$

where A_i = the virtual strain of the ith node in any four-node element,

$$\left(Lu - f, \sum_{i=1}^{4} N_i A_i\right) = 0 \tag{5.16}$$

Because of the randomicity of A_i, we can obtain $(Lu, f, N_i) = 0, i = 1, 2, 3, 4$, namely,

$$\iint_e N_i (Lu - f) \, dx \, dy = 0$$

5.3.2.2 Fundamental Equation Set

Through weighted residual method and Galerkin method, we can transform weight function into shape function, and typical equation set can be obtained from formula (5.8) and formula (5.11)

$$\iint_A N_i \left\{ \left[(\lambda+2G)\frac{\partial^2}{\partial x^2} + G\frac{\partial^2}{\partial y^2}\right] u + (\lambda+G)\frac{\partial^2 v}{\partial x \partial y} - \frac{\partial p}{\partial x} - \frac{\partial u_g}{\partial x} - \rho \ddot{u} \right\} dx \, dy = 0$$

$$\tag{5.17}$$

5.3 Theoretical Analysis

$$\iint_A N_i \left\{ \left[G \frac{\partial^2}{\partial x^2} + (\lambda+2G) \frac{\partial^2}{\partial y^2} \right] v + (\lambda+G) \frac{\partial^2 u}{\partial x \partial y} - \frac{\partial p}{\partial y} - \frac{\partial u_g}{\partial y} - \rho \ddot{v} + \rho g \right\} dxdy = 0$$

(5.18)

$$\iint_A N_i \left[-\frac{\partial}{\partial t} \left(\frac{\partial u}{\partial x} + \frac{\partial v}{\partial y} \right) + \overline{k_x} \frac{\partial^2 p}{\partial x^2} + \overline{k_y} \frac{\partial^2 p}{\partial y^2} \right] dxdy = 0 \quad (5.19)$$

$i = 1, 2, 3, \ldots, n$; A denotes the definition domain of unknown functions.

For avoiding that the second derivatives of functions above are infinite, the integral requires the slopes of all the interface area are continuous. For eliminating this limitation, integration by parts is adopted for transformation, such as

$$\iint_A N_i \frac{\partial^2 u}{\partial x^2} dxdy \equiv \int \left| N_i \frac{\partial u}{\partial x} \right| dy - \iint_A \frac{\partial N_i}{\partial x} \frac{\partial u}{\partial x} dxdy \equiv \oint_S N_i \frac{\partial u}{\partial x} l_x ds - \iint_A \frac{\partial N_i}{\partial x} \frac{\partial u}{\partial x} dxdy$$

(5.20)

where

l_x = the x direction cosine of the boundary's exterior normal
S = the path of integration, which is the boundary lines

After applying this method, formulas (5.17), (5.18), and (5.19) are transformed into

$$\iint_A \left[(\lambda+2G) \frac{\partial N_i}{\partial x} \frac{\partial u}{\partial x} + G \frac{\partial N_i}{\partial y} \frac{\partial u}{\partial y} + \left(\lambda \frac{\partial N_i}{\partial x} \frac{\partial v}{\partial y} + G \frac{\partial N_i}{\partial y} \frac{\partial v}{\partial x} \right) - \frac{\partial N_i}{\partial x} (p+u_g) \right] dxdy =$$

$$- \iint_A N_i \rho \ddot{u} dxdy + \oint_S N_i \left[(\lambda+2G) \frac{\partial u}{\partial x} l_x + G \frac{\partial u}{\partial y} l_y + \lambda \frac{\partial v}{\partial y} l_x + G \frac{\partial v}{\partial x} l_y - (p+u_g) l_x \right] ds$$

(5.21)

$$\iint_A \left[G \frac{\partial N_i}{\partial x} \frac{\partial u}{\partial y} + \lambda \frac{\partial N_i}{\partial y} \frac{\partial u}{\partial x} + G \frac{\partial N_i}{\partial x} \frac{\partial v}{\partial x} + (\lambda+2G) \frac{\partial N_i}{\partial y} \frac{\partial v}{\partial y} - \frac{\partial N_i}{\partial y} (p+u_g) \right] dxdy =$$

$$\iint_A N_i \rho (-\ddot{v} + g) dxdy + \oint_S N_i \left[\lambda \frac{\partial u}{\partial x} l_y + (\lambda + 2G) \frac{\partial u}{\partial y} l_x + G \frac{\partial v}{\partial x} l_x \right.$$

$$\left. + (\lambda + 2G) \frac{\partial v}{\partial y} l_y - (p + u_g) l_y \right] ds \quad (5.22)$$

$$\iint_A \left[-N_i \frac{\partial}{\partial x} \frac{\partial u}{\partial t} - N_i \frac{\partial}{\partial y} \frac{\partial v}{\partial t} - \left(\bar{k}_x \frac{\partial N_i}{\partial x} \frac{\partial}{\partial x} + \bar{k}_y \frac{\partial N_i}{\partial y} \frac{\partial}{\partial y} \right) p \right] dxdy =$$

$$- \oint_S N_i \left(\bar{k}_x l_x \frac{\partial}{\partial x} + \bar{k}_y l_y \frac{\partial}{\partial y} \right) p ds \qquad (5.23)$$

Substituting u, v, and p by the above approximate formulas (5.12) and taking the geometric formulas and constitutive relations into account, we can obtain

$$\sum_{j=1}^{n} \left[m_{ij} \ddot{u}_j(t) + K_{ij}^{11} u_j(t) + K_{ij}^{12} v_j(t) + K_{ij}^{13} p_j(t) \right] = F_i^1(t) \qquad (5.24)$$

$$\sum_{j=1}^{n} \left[m_{ij} \ddot{v}_j(t) + K_{ij}^{21} u_j(t) + K_{ij}^{22} v_j(t) + K_{ij}^{23} p_j(t) \right] = F_i^2(t) \qquad (5.25)$$

$$\sum_{j=1}^{n} \left[K_{ij}^{31} u_j(t) + K_{ij}^{32} v_j(t) + K_{ij}^{33} p_j(t) \right] = F_i^3(t) \qquad (5.26)$$

where

$$i = 1, 2, 3, 4, \ldots, n$$

$$m_{ij} = \iint_A \rho N_i N_j dxdy$$

$$K_{ij}^{11} = \iint_A \left[(\lambda + 2G) \frac{\partial N_i}{\partial x} \frac{\partial N_j}{\partial x} + G \frac{\partial N_i}{\partial y} \frac{\partial N_j}{\partial y} \right] dxdy$$

$$K_{ij}^{22} = \iint_A \left[G \frac{\partial N_i}{\partial x} \frac{\partial N_j}{\partial x} + (\lambda + 2G) \frac{\partial N_i}{\partial y} \frac{\partial N_j}{\partial y} \right] dxdy$$

$$K_{ij}^{23} = -\frac{1}{2} \iint_A \left[\bar{k}_x \frac{\partial N_i}{\partial x} \frac{\partial N_j}{\partial x} + \bar{k}_y \frac{\partial N_i}{\partial y} \frac{\partial N_j}{\partial y} \right] dxdy$$

$$K_{ij}^{12} = \iint_A \left[\lambda \frac{\partial N_i}{\partial x} \frac{\partial N_j}{\partial y} + G \frac{\partial N_i}{\partial y} \frac{\partial N_j}{\partial x} \right] dxdy$$

$$K_{ij}^{21} = \iint_A \left[G \frac{\partial N_i}{\partial x} \frac{\partial N_j}{\partial y} + \lambda \frac{\partial N_i}{\partial y} \frac{\partial N_j}{\partial x} \right] dxdy$$

5.3 Theoretical Analysis

$$K_{ij}^{13} = -\iint_A N_j \frac{\partial N_i}{\partial x} dxdy$$

$$K_{ij}^{31} = -\iint_A N_i \frac{\partial N_j}{\partial x} dxdy$$

$$K_{ij}^{23} = -\iint_A N_j \frac{\partial N_i}{\partial y} dxdy$$

$$K_{ij}^{32} = -\iint_A N_i \frac{\partial N_j}{\partial y} dxdy$$

$$F_i^1(t) = \iint_A N_i \frac{\partial N_i}{\partial x} u_g dxdy - \oint_S N_i \left(\sigma_x l_x + \tau_{xy} l_y + p l_x + u_g l_x\right) ds$$

$$F_i^2(t) = \iint_A N_i \left(\rho g + \frac{\partial N_i}{\partial y} u_g\right) dxdy - \oint_S N_i \left(\sigma_y l_y + \tau_{xy} l_x + p l_y + u_g l_y\right) ds$$

$$F_i^3(t) = -\oint_S N_i \left(q_x l_x + q_y l_y\right) ds$$

where $\sigma_y l_y + \tau_{xy} l_x + p l_y + u_g l_y = f_y$ is the boundary load, which is known as the dynamic load of trains, and $q_x l_x + q_y l_y = q_n$ is the known seepage capacity of boundary.

We can take the derivative of formulas (5.24), (5.25), and (5.26) with respect to time t and assume the increment of the displacement (u_i, v_i) and pore water pressure (p_i) of node i in the period of Δt as Δu_i, Δv_i. Meanwhile, Δp_i, u_{io}, v_{io}, and p_{io} are given as initial values. The mean value p_i of this period is $p_{io} + \Delta p_i/2$, and the formula (5.24), (5.25), and (5.26) can be written as the following difference equations:

$$\sum_{j=1}^{n} \left[m_{ij} \Delta \ddot{u}_j + K_{ij}^{11} \Delta u_j + K_{ij}^{12} \Delta v_j + K_{ij}^{13} \Delta p_j\right] = \Delta F^1 \quad (5.27)$$

$$\sum_{j=1}^{n} \left[m_{ij} \Delta \ddot{v}_j + K_{ij}^{21} \Delta u_j + K_{ij}^{22} \Delta v_j + K_{ij}^{23} \Delta p_j\right] = \Delta F^2 \quad (5.28)$$

$$\sum_{j=1}^{n} \left[K_{ij}^{31} \Delta u_j + K_{ij}^{32} \Delta v_j + K_{ij}^{33} \Delta p_j\right] = \Delta \overline{F^3} \quad (5.29)$$

$$\Delta \overline{F^3} = \Delta F_i^3 - 2\Delta t \sum_{j=1}^{n} K_{ij}^{33} p_{j0} \quad (5.30)$$

Because the gravity (ρg) doesn't vary with time, the right item of formulas (5.27), (5.28), and (5.29) can be determined through the increment of load, pore water pressure, and seepage volume:

$$\Delta F_i^1(t) = \iint_A \frac{\partial N_i}{\partial x} \Delta u_g \, dx \, dy - \oint_S N_i \Delta f_x \, ds \qquad (5.31)$$

$$\Delta F_i^2(t) = \iint_A \frac{\partial N_i}{\partial y} \Delta u_g \, dx \, dy - \oint_S N_i \Delta f_y \, ds \qquad (5.32)$$

$$\Delta F_i^3(t) = -\oint_S N_i \Delta q_n \, ds \qquad (5.33)$$

where

Δf_x, Δf_y, Δq_n = the known conditions of the boundary
Δu_g = the dynamic pore water pressure generated in Δt time

The equation set consists of $3n$ linear algebraic equations with $3n$ unknowns; for any moment, it can be solved by triangular decomposition.

5.3.3 Static-Dynamic Analysis of Subway-Soil System

Applying the above theory into the dynamic response analysis of soil-tunnel-train system, the right items of the formulas (5.27) and (5.28) contain the pore water pressure increment Δu_g induced by the vibration of subway. This pore water pressure increment can only be obtained by dynamic analysis. Considering the nonlinearity of shear modulus G and damping coefficient D, we adopted iteration method in computation. While the soil is under dynamic loading, the pore water pressure will increase, diffuse, and dissipate simultaneously. Because the dynamic pore water pressure increment is crucial to the whole computation, the whole dynamic analysis must be conducted by stages. Through piecewise iteration, pore water pressure increment Δu_g during each stage can be carried out; then transform Δu_g into nodal loading and substitute it for the ΔF in equations (5.27) and (5.28); finally, the displacement (u, v) and residual pore water pressure (p) of each node can be figured out. Meanwhile, since the displacement is known, the strain ε and effective stress σ can be obtained. According to the mean effective stress σ_m, a new shear modulus G can be calculated which will be employed in the next iteration computation. Similarly, the new damping coefficient can be obtained by the mean shear strain of the former stage. The iteration will not be over until the last stage finishes. More specifically:

5.3 Theoretical Analysis

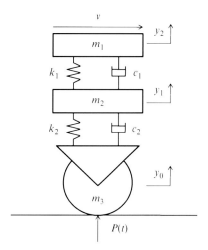

Fig. 5.1 Simplified model for the interaction in a wheel-rail system

1. The static computation is conducted; the nonlinear static stress and stress level in each element can be obtained through static FEM and Duncan-Chang secant modulus formula.
2. Considering the nonlinear properties of soil, Hardin-Drnevich model is applied in the calculation of dynamic shear modulus G and damping coefficient D.
3. Applying step-by-step integration to solve the kinematical equilibrium equation below:

$$\underline{M}\ddot{U} + \underline{C}\dot{U} + \underline{K}U = F(t) \tag{5.34}$$

where

$\underline{M}, \underline{C}, \underline{K}$ = the integral mass matrix, damping matrix, and stiffness matrix of the system
$U, \dot{U}, \ddot{U}$ = the displacement, velocity, and acceleration of each node relative to the bottom of the tunnel
$F(t)$ = the dynamic loading induced by train which is applied on the node of the rail

The specific solving processes of $F(t)$ are listed as follows:

In Fig. 5.1, m_1 denotes the sprung mass; m_2 and m_3 denote the unsprung mass; k_1, k_2 denote spring stiffness; c_1, c_2 denote damping; $P(t)$ denotes the interaction force between wheel and rail; and y_0, y_1, and y_2 denote the vertical coordinate of m_1, m_2, and m_3. The kinematical equilibrium equation of the wheel-rail system in Fig. 5.1 can be written as

$$m_1 \ddot{y}_2 + k_1 (y_2 - y_1) + c_1 (\dot{y}_2 - \dot{y}_1) = 0$$
$$m_2 \ddot{y}_1 + k_2 (y_1 - y_0) + c_2 (\dot{y}_1 - \dot{y}_0) + k_1 (y_1 - y_2) + c_1 (\dot{y}_1 - \dot{y}_2) = 0 \quad (5.35)$$

where $\dot{y} = dy/dt$, $\ddot{y} = d^2y/dt^2$

The interaction between the wheel and rail is considered as complete contact; the vertical acceleration of the wheel-rail system $\ddot{y}_0$ is assumed equal to the acceleration of the rail's bottom, namely,

$$\ddot{y}_0 = \sum_{n=0}^{\frac{n}{2}-1} (A_n \cos nwt + B_n \sin nwt) \quad (5.36)$$

According to the dynamic equilibrium condition in the vertical direction, the interaction force of wheel and rail can be obtained:

$$P(t) = m_1 \ddot{y}_2 + m_2 \ddot{y}_1 + m_3 \ddot{y}_0 + (m_1 + m_2 + m_3) g \quad (5.37)$$

If $P(t)$ is assumed distributes along the track uniformly, it can be translated into equivalent linear load $f(t)$, which is shown in Eq. (5.38)

$$f(t) = \frac{P(t)}{L_1} \times n_1 \quad (5.38)$$

where

$n_1 = $ the number of wheel sets of one carriage, which is 4
$L_i = $ the length of one carriage, which is 20.368 m

In the computation of damping matrix, Rayleigh damping is adopted:

$$\underline{C} = \alpha \underline{M} + \beta \underline{K} \quad (5.39)$$

where α, β = damping coefficient. According to the empirical method, $\alpha = 0.03$, $\beta = 0.01$.

Newmark implicit time integration method is employed to solve the dynamic equation (5.34) of this system. The analytical process should obey the Mohr-Coulomb yield criterion and associated flow rule, and the stress is required to be adjusted if it is over the yield stress. The displacement, velocity, and acceleration of each node as well as the stress and strain of each element in a period can be figured out through numerical integration. The new shear modulus G_i can be obtained through substituting mean dynamic shear strain γ_m in formula (5.2). If the new shear modulus G_i and the initial shear modulus for the iteration $G_{i,1}$ meet the convergence criteria, the iteration will stop. Otherwise, it will restart calculation till they meet the requirement.

4. According to formula (5.1), the pore water pressure increment Δu_g of each element in a period can be calculated. Then, ΔF^1 and ΔF^2 can be obtained. Finally, we can get the displacement u, v, and residual pore water pressure p of each node.
5. Based on the resultant dynamic strain and mean effective stress, the shear modulus after the increment of pore water pressure can be calculated for the next stage of iteration. Similarly, the new damping coefficient can be worked out for the next iteration as well.
6. Repeat the steps above until the computational vibration time reaches the preset value.

5.4 Simulation of the Dynamic Loading Induced by the Subway Train

The promotion of the rapid economic growth and intense social demands spurs many urban cities to construct their subway system in China. Though the subway has played an important role, a series of problems caused by the operation of subway system gradually emerge, such as the environmental vibration, the noise, and the displacement of the structures near the subway system. For studying these problems, the determination of dynamic load induced by subway trains is the precondition. Unfortunately, the simulation of subway loading involves the study of the dynamic response of a complicated system which consists of train, rail and ballast bed, and even the foundation. In the study, numerous subjects are involved, including vehicle dynamics, orbits dynamics, wheel-rail interaction, soil dynamics, and structural dynamics. However, for a general study, the simplified model of the train-rail system can already meet the accuracy requirement. A precise model which can simulate the whole system and taking the nonlinear properties of interaction into account is unnecessary.

Pan and Xie (1990), Gao (1998), Li (1998), Zhang and Bai (2000), and Wang et al. (2005) simulated the dynamic load on the wheel-rail interaction basing on the rail acceleration obtained from the field test. Feng et al. (2007; Feng and Yan 2008) applied the rail acceleration obtained from the field test to determine the dynamic load and compared the results with the dynamic load determined through track irregularity. These two methods all adopt frequency-domain analysis (discrete Fourier transform). Frequency-domain analysis transforms the tested acceleration into the frequency-domain for obtaining the response solution through the determination of transformation coefficient. The dynamic load (response) in time domain can be obtained through inverse Fourier transform. Because the damping orthogonal assumption and superposition principle are introduced in the solving process, this method can only be applied in the study of linearly systematic problems.

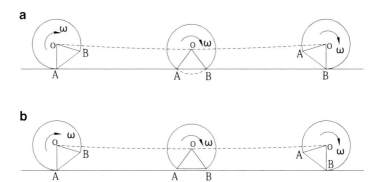

Fig. 5.2 (**a**) The motion of the wheel with flat under low speed and (**b**) the motion of the wheel with flat under high speed

5.4.1 Generation Mechanism

The subway tunnel is a semi-enclosed space; when the subway trains are running, the interaction force between the wheel and the rail is the primary dynamic source for the vibration of tunnel system. This interaction force can be affected by the defects of wheels and track. For the wheels, the influence factors primarily include the wheel flat and eccentric wheel. For the track, the dominant factor is the geometric track irregularity.

5.4.1.1 Defective Wheel

Wheel Flat

For a long-term moving subway train, the tread of wheels tend to have partial abrasion and stripping due to the braking and idling, which are all named as wheel flat. The wheel flat can induce special kinetic performance in the system when the wheels are rotating (Fig. 5.2).

Eccentric Wheel

If the centroid of the wheel is not coincident to the geometric center, which means a small eccentric distance exists in the wheel, a constant and unbalanced inertia force F will be generated in the operation of the train (Fig. 5.3).

$$F = M\omega^2 r_0 = M\left(\frac{V}{R}\right)^2 r_0 \tag{5.40}$$

5.4 Simulation of the Dynamic Loading Induced by the Subway Train

Fig. 5.3 The motion of an eccentric wheel

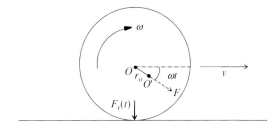

where

M = mass of the wheel
ω = angular velocity
V = speed of the train
R = radius of the wheel

The vertical component of the inertia force F is the periodic vibrating load generated by eccentric wheel, which can be expressed as

$$F_v(t) = M\left(\frac{V}{R}\right)^2 r_0 \sin\left(\frac{V}{R}t\right) \qquad (5.41)$$

5.4.1.2 Track Irregularity

Track is a special kind of structure. When the track undergoes intense cyclic random dynamic load for a long period, the uneven residual deformation of the track will appear, so the geometric state of track is constantly changing along the route. Besides, the inevitable errors in construction can lead to the track irregularity as well.

When a subway train is passing a completely even track, a uniform displacement y_0 will be generated. In this case, there is no additional displacement and dynamic load for the rail and wheels. However, in the irregular zone of the rail, a new dynamic equilibrium condition is developed. The subway train will cause forced vibration in this zone, which will induce the additional displacement of the rail and additional dynamic loading to the wheels (Luo and Gen 1999). Before the running train enters the irregular zone, the centroid of the wheel is parallel to the track. Once the wheel reaches the irregular zone, the wheels may plunge for the depth of y_h which will bring forced vibration into the wheel-rail system and lead to additional displacement y_b of the track. This forced vibration and additional displacement will last until the wheel passes the terminal point of the irregular section.

Lengthways Height Variation

The height variation of lengthways tracks can be generated by many factors, like rail abrasion, nonuniform elastic deformation of track components (elastic cushion, sleeper, track bed, subgrade), and small gaps between the components. Though the existence of rail joints can cause vertical vibration as well, most subway systems adopt seamless technology currently which can overcome this problem. The vertical profile irregularity is the primary factor of the vertical vibration of the subway train. It can lead to massive inertia force. The irregularity in long wavelength mainly affects the vibration of cars (sprung mass), while the irregularity of short wavelength can cause larger interaction force and unsprung acceleration.

Horizontal Height Difference

The horizontal height difference denotes the height difference between the right and left wheel; it is the main reason for the side-rolling of trains. If the amplitude of the horizontal height difference is large and continuous, when the frequency of the train's side-rolling approaches the natural frequency of the train, the train can demonstrate intense rocking vibration from side to side. This vibration causes the increase of load on one side, while the decrease on the other side, which may lead to train derailment.

Track Axis Variation

The track axis variation can be induced by many factors, like the uneven abrasion of rail profiles, varying transverse elasticity and damping, and failed rail fasteners. The track axis variation can lead to the traverse-rolling vibration which can generate intense lateral force and roll torque in train-track system, especially in the hunting motion of trains. If the intense lateral force excesses the allowable value, it may threat the normal operation of trains.

Track Gauge Deviation

The track gauge deviation denotes the difference between the real gauge and standard gauge; it primarily affects the lateral stability of the running train and rail/wheel interface abrasion. The oversize distance may lead to the derailment (Liang and Cai 1999).

5.4 Simulation of the Dynamic Loading Induced by the Subway Train

Fig. 5.4 Schematic diagram of a subway train

5.4.2 Simplified Model for Subway Train

The subway train consists of six cars which contain two trailers and four motor cars. The schematic diagram of the train is shown in Fig. 5.4. *A* denotes the trailers which are equipped with the cabs; *B* denotes the motor cars which are equipped with pantographs; *C* denotes the motor cars; and *A*, *B*, and *C* compose an independent train unit. The joints between the cars adopt the energy-absorbing coupler which is a kind of automatic car coupler. The subway train used in Shanghai is made in Germany. Each car of subway train mainly consists of the car bodies, car couplers, bogies and some other components. Among them, the bogie is one of the most crucial components; it contains wheel sets, axle box, frameworks, spring suspension, absorbers, and so on.

Because the vertical vibration of subway trains is commonly considered as more influential on the tunnel structure than the rolling and lateral vibration, we take the vertical vibration as the focus of the study. Considering the car is usually symmetrical in lateral direction, the bounce effect on the rail-wheel interaction and the sway of car bodies can be ignored. We assumed the weight of the whole train is distributed to each wheel set evenly and all the wheels, bogies, and car bodies are considered as rigid.

The frequencies of the subway train load can be divided into three categories: high frequency, middle frequency, and low frequency. According to the data obtained from the field test, the load of middle and high frequency was absorbed rapidly by the tunnel lining and vibration absorption; the dynamic load which affected the soil around the tunnel was primarily made of the low-frequency load. Therefore, in the numerical simulation conducted in this chapter, the simulated load was the low-frequency load, which was generated by the wheel-rail interaction force. This load would affect the soil around the tunnel through the track system and tunnel lining.

5.4.3 The Simulation of Subway Train Loading

According to the method introduced above, we simulated the subway train load through the program executed in ANSYS Parametric Design Language (APDL)

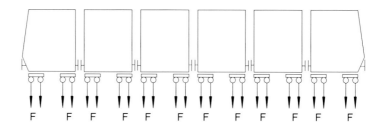

Fig. 5.5 Schematic diagram of subway train load

(shown as below). Assuming the passengers in each car reach the transport capacity, we set the weight of the subway train as 333.46 tons; the load is distributed averagely to each wheel set. The simulation of the dynamic load is shown in Fig. 5.5.

```
*del,all
rlen=300*1000              ! Define the track length
rcount=100                 ! Define the layer amount of
                             track elements
raxi='y'                   ! Define coordinate axis of
                             the track in coordinate
                             system
rdir=1                     ! Define the sign of the track
                             on the coordinate axis
taxi='z'                   ! Define coordinate axis
                             of the weight of the train
                             in coordinate system
tdir=1                     ! Define the sign of the
                             weight of the train on the
                             coordinate axis
load_com='load'            ! Define the name of the nodes
                             on the track surface
tvel=45*1000*1000/60/60    ! Define the speed of the
                             train
tnum=6                     ! Define the amount of the
                             cars
......
************
time_scale=1               ! Define the amplification
                             coefficient of time step
*if,tggslength,ge,rlen/rcount,then
......
         tggscount=nint(tggslength/(rlen/rcount))+1
```

5.4 Simulation of the Dynamic Loading Induced by the Subway Train

```
            *else
            tggscount=nint(tggslength/(rlen/rcount))
         *end if
*else
         tggscount=1
*end if
*if,tggllength,ge,rlen/rcount,then
......
            tgglcount=nint(tggllength/(rlen/rcount))+1
*else
            tgglcount=nint(tggllength/(rlen/rcount))
*end if
*else
tgglcount=1
*end if
*if,trrlength,ge,rlen/rcount,then
......
            trrcount=nint(trrlength/(rlen/rcount))+1
*else
            trrcount=nint(trrlength/(rlen/rcount))
*end if
*else
trrcount=1
*end if
tload=tgra/tnum/2/4
time_each=rlen/rcount/tv  ! Define the time required to
                            pass two adjacent nodes for
                            the train
time_total=time_each*(rcount+1)+(tggscount*tnum*2+
            tgglcount*tnum+trrcount*(tnum.1))
*time_each                  ! Define the time required to
                            pass the track completely
                            for the train
*dim,road_load,Table ,time_total/time_each+1,rcount+1,
            1,time,%raxi%,,
! Define the array used for depositing the track load
*dim,train_load,Table ,tggscount*tnum*2+tgglcount*tnum
            +trrcount*(tnum.1)+1,,,
! Define the array used for depositing the train load
```

```
*do,i,1,time_total/time_each+1,1  road_load(i,0)=(i.1)*
         time_each*enddo
             ! Define the timeline of the track array
*do,i,1,rcount+1,1  road_load(0,i)=(i.1)*(rlen/rcount)*
         rdir*enddo
       ! Define the coordinate axis of the track array
*do,i,1,tggscount*(tnum*2.1)+tgglcount*tnum+trrcount*
         (tnum.1)+1,tggscount*2+tgglcount+trrcount
train_load(i,1)=tload*tdir   !train_load(i,0)=(i.1)*
time_each
! Only applied for checking
......
  train_load(i+tggscount+tgglcount+tggscount,1)=
tload*tdir
  !train_loadi+tggscount+tgglcount+tggscount,0)=
(i+tggscount+tgglcount+tggscount.1)
*time_each
! Only applied for checking
*enddo
!Attach the value to the train load
*do,i,1,rcount+1,1*vfun,road_load(i+1,i),copy,
train_load(1,1)*enddo/solcmsel,s,%load_com%
f,all,f%taxi%,%road_load%
! Apply load
!***************solution********************
antype,4
trnopt,full
lumpm,0
deltim,time_each*time_scale,0,0
autots,off
time,nint(time_total/(time_each*time_scale))*time_each
*time_scale
eqslv,pcg,1e.6
outres,erase
outres,all,1
rescontrl,define,all,1,1
allsel
eplot
!/input,output,mac
! Only applied for checking
```

5.5 Model Development

5.5.1 Introduction to the Model

The prototype of this numerical model is the subway tunnel between Jing'An Temple Station and Jiang Su Road Station of Line 2 in Shanghai Metro. According to the geological survey of the field (Table 5.1), the stratigraphic profile of the model is shown in Fig. 5.6.

According to the engineering geology profile in the field survey data, we input the borehole data in ANSYS and connect the layer break point of two adjacent boreholes by splines. Then, we extended splines of the same stratum to curved surfaces which can be treated as the interfaces of stratums (Figs. 5.7 and 5.8). Finally, the solids which are applied in the simulation of corresponding stratums can be generated by the curve surfaces. Since the focus of this study is the soil around the tunnel, we

Table 5.1 Depth of the stratums

Strata serial number	Strata name	Depth (m)
②	Isabelline silty clay	2.5–3.6
③	Gray mucky silty clay	6.8–7.6
④	Gray mucky clay	15.2–16.8
⑤1a	Gray clay	19.0–20.2
⑤1b	Gray silty clay	27.0–28.0
⑤1c	Gray mucky clay/sandy silt	37.0–38.8

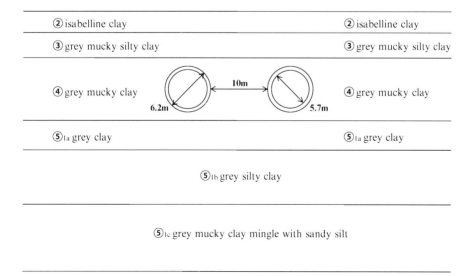

Fig. 5.6 Stratigraphic profile of the soil model

Fig. 5.7 Strata generation

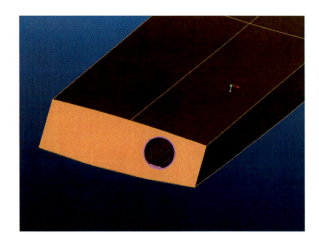

Fig. 5.8 Strata interfaces

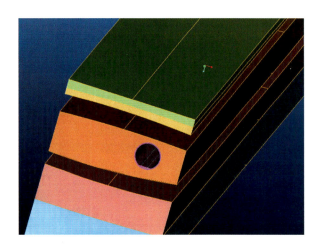

assumed the subway tunnel is constructed horizontally; the rails are constructed on the bottom of the tunnel lining (Fig. 5.9). The depth of the tunnel axis is set to be 11.5 m, the inner diameter of the tunnel is 5.7 m, and external diameter is 6.2 m. The final developed model is shown in Fig. 5.10.

According to the field survey data and our computing experience, the results can meet the accuracy requirement by calculating half of the model, which can reduce the computational duration largely. Along the depth, since the strength of sixth stratum which consists of silty clay is high, the displacement is quite small. The displacement is generated from the second stratum to the fifth stratum primarily. In the computation, the calculated depth is set for 38 m which is the bottom of the fifth stratum. If the outcomes of the strain excess the allowable value in this depth, we could extend the computation depth in this model.

5.5 Model Development

Fig. 5.9 Tunnel lining and rail

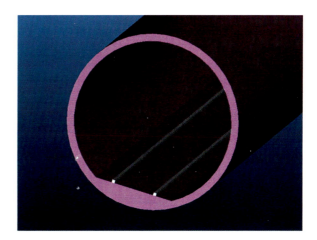

Fig. 5.10 Geometric model

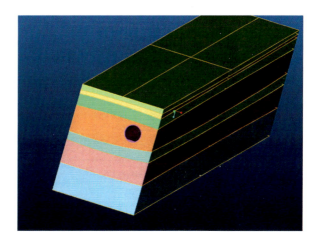

5.5.2 Mesh Generation in FEM Model

For the dynamic numerical analysis of a semi-infinite space, the numerical result can only be considered as accurate when the generated mesh can simulate the wave shape (Huang et al. 1999). In the physical perspective, the discretization of continuous medium can lead to two adverse affects. One is called low-pass effect, and the other is called dispersion effect; they are capable to change the properties of wave propagation. Although these errors are inevitable, they can be ignorable when the size of the mesh is small enough, which has been proved by the theory and practice. In the two-dimensional and three-dimensional discrete models, more problems appeared besides the frequency dispersion and cutoff frequency in one-dimensional discrete model. Therefore, when we are analyzing the propagation of

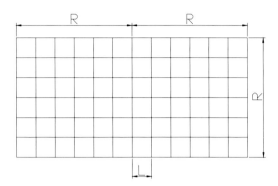

Fig. 5.11 Example of a meshed finite element model

the transient waves, the size of the mesh should be small enough. As an example, the mesh generation of a semi-infinite space model is conducted, this model is applied with a unit harmonic loading.

In Fig. 5.11, R denotes the distance between the vibration source and the boundary, and L denotes the length of a finite element. There are two requirements in the mesh generation:

5.5.2.1 The Size of the Mesh R

The size of the mesh should be large enough to obtain a fixed displacement of a point. The trial computation demonstrated that the results of displacement had converged well when $R = 1.5\lambda_s$, while the results showed more errors when $R = 0.5\lambda_s$ (λ_s denotes the wavelength of the shear wave).

5.5.2.2 The Length of an Element

When $L = \lambda_s/14$, the outcome of the whole numerical model is accurate. When $L = \lambda_s/6$, the outcome of the numerical model is partially accurate. For the elements which are apart from the source for $0.5\lambda_s$, the outcomes are accurate, while for the elements which are closed to the source, the outcomes are not satisfying. However, these errors can be eliminated by reducing the length of elements. Based on the trial computation, two conclusions can be obtained: first, the size of the FEM model should be over $1.5\lambda_s$; second, the length of the elements should be less than $\lambda_s/14$. The length of the waves λ_s depends on the frequency:

$$\lambda_s = 2\pi \cdot V_s/\omega = V_s/f \tag{5.42}$$

where V_s denotes the shear wave velocity of soil and f denotes frequency. For layered soil, the shear wave velocities of different layers are different. Therefore,

5.5 Model Development

Table 5.2 Results of wave velocity field test

Strata serial number	Strata name	Velocity of shear wave (m/s)	Velocity of compression wave (m/s)
②	Isabelline silty clay	123	1,079
③	Gray mucky silty clay	119	1,139
④	Gray mucky clay	120	1,208
⑤₁ₐ	Gray clay	145	1,267
⑤₁ᵦ	Gray silty clay	199	1,284
⑤₁꜀	Gray mucky clay/sandy silt	241	1,312
⑥	Dark green silty clay	268	1,355

for obtaining the shear wave velocities of the test site, we conducted wave velocity field test, and the result is shown in Table 5.2.

Because the subway tunnel of Shanghai is mainly constructed in the fourth soil layer, the shear wave velocity applied in the numerical model is the data of the fourth soil layer, which is 120 m/s. Cheng (2003) concluded the predominant period in Shanghai varies between 0.35 and 2.40 s; therefore, the frequency can be taken between 0.5 and 3.0 Hz. According to the formula (5.42), the shear wave length λ_s is varying between 40 and 240 m. Basing on the above conclusions, the parameters applied in the mesh generation were determined: the distance between vibration source and boundary R is over 360 m; the length of elements which are near the source is less than 2.9 m, while the length of elements which are far from the source is less than 6.7 m. For meshing the model smoothly, node bias was applied. There are three methods for setting the bias:

5.5.2.3 Linear Bias

In the linear bias method, the degree of bias is related to the positive slope of a line, given

$$s \in \left\{0, \frac{1}{n}, \frac{2}{n}, \frac{3}{n}, \ldots, \frac{n}{n}\right\} \tag{5.43}$$

$$x(s) = x(0) + C \int_0^s (m\varepsilon + b) \, d\varepsilon = x(0) + C\left(\frac{ms^2}{2} + bs\right) \tag{5.44}$$

where n is defined as the element density, $x(s)$ ($0 \leq s \leq 1$) denotes a function for node placement, m denotes the slope of the line, and b denotes the intercept of the line.

Substituting $x(0) = 0$ and $x(1) = 1$ into Eq. (5.44), $C = 2/(m + 2b)$ can be obtained so

$$x(s) = s\frac{ms + 2b}{m + 2b} \tag{5.45}$$

where m reflects the bias degree, b is applied as proportionally coefficient to enlarge or reduce the influence of m, and the value of b applied in this model is 1.5.

5.5.2.4 Exponential Bias

In this method, the intervals are varying in geometrical progression. Given

$$s \in \left\{0, \frac{1}{n}, \frac{2}{n}, \frac{3}{n}, \ldots, \frac{n}{n}\right\} \tag{5.46}$$

n is defined as element density, $x(s)$ is given as a function for node placement, and the value of the function is varying from 0 to 1: $x(0) = 0$, $x(1) = 1$.

$$x(s) = \frac{\int_0^s \phi(\varepsilon)\,d\varepsilon}{\int_0^1 \phi(\varepsilon)\,d\varepsilon} \tag{5.47}$$

Given $\phi(\varepsilon) = C^{n\varepsilon}$, for

$$\phi(0) = 1, \quad \phi(1/n) = C, \quad \phi(2/n) = C^2, \quad \phi(3/n) = C^3, \ldots, \phi(1) = C^n$$

where $C = 1.0 + 0.1abs(\text{bias\_intensity})$ is the geometric growth coefficient and $x(s)$ can be rewritten as

$$x(s) = \frac{C^{ns} - 1}{C^n - 1} \tag{5.48}$$

5.5.2.5 Symmetrical Exponential Bias

In this method, nodes are central symmetrically distributed on the model side. If the bias density is positive, the small intervals are placed at the two ends of the side; if the value is negative, the small intervals are placed at the central part of the side. Given

$$s \in \left\{0, \frac{1}{n}, \frac{2}{n}, \frac{3}{n}, \ldots, \frac{n}{n}\right\} \tag{5.49}$$

5.5 Model Development

Fig. 5.12 Diagram of mesh generation

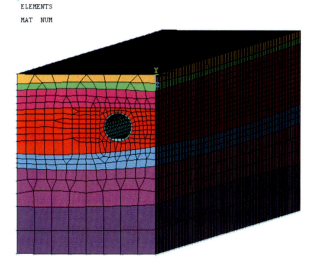

n is defined as the element density, $x(s)$ is given as a function for node placement, and the value of the function is varying from 0 to 1: $x(0) = 0$, $x(1) = 1$.

$$x(s) = \frac{1}{\int_0^1 \phi(\varepsilon)d\varepsilon} \int_0^s \phi(\varepsilon)\,d\varepsilon \tag{5.50}$$

Given $\phi(\varepsilon) = e^{-r(\varepsilon - 0.5)^2}$, $x(s)$ can be transformed into

$$x(s) = \frac{\operatorname{erf}\left(\frac{\sqrt{r}}{2}\right) - \operatorname{erf}\left(\sqrt{r}\left(\frac{1}{2} - s\right)\right)}{2\operatorname{erf}\left(\frac{\sqrt{r}}{2}\right)} \tag{5.51}$$

$$\operatorname{erf}(t) = \int_0^t e^{-r(\varepsilon - 0.5)^2}\,d\varepsilon \tag{5.52}$$

where $\operatorname{erf}(t)$ denotes the statistical Gauss error function.

The model through the mesh process according to the method referred above is demonstrated in Fig. 5.12.

Table 5.3 Physical and mechanics parameters of soil

Strata serial number	Strata Name	C (kPa)	φ (°)	$a_{0.1-0.2}$ (MPa^{-1})	$E_{s0.1-0.2}$ (MPa)	γ (kN/m^3)
②	Isabelline silty clay	17	16.4	0.46	4.39	18.4
③	Gray mucky silty clay	4	29.7	0.74	3.03	17.5
④	Gray mucky clay	11	23.4	1.19	2.07	16.9
⑤1a	Gray clay	32	18.4	0.72	3.13	18.0
⑤1b	Gray silty clay	36	30.5	0.48	4.24	18.1
⑤1c	Gray mucky clay/sandy silt	51	29.5	0.41	4.88	18.2

Table 5.4 Material parameters of subway lining and rail

Material	Poisson's ratio μ	Elastic modulus E (MPa)	Unite weight γ (kN/m^3)
Tunnel lining	1/6	35.5	25

5.5.3 Material Parameters

The selection of the material parameters in the numerical model is crucial. It could determine the final results directly. In this FEM model, the parameters which are shown in Tables 5.3 and 5.4 are all obtained from the field investigation and tests.

5.6 Viscoelastic Artificial Boundary

For the dynamic analysis, the geological medium can be considered as semi-infinite media. The stress waves propagate from the tunnel to the surrounding soil. For obtaining accurate results without an overlarge computational domain, the artificial boundary is applied in the discretization of the numerical model.

When finite element method is applied to analyze the dynamic interaction of the structure and soil, finite computational domain must be taken out from the semi-infinite geological medium. On the boundary of the taken-out domain, artificial boundary is required for simulating the radiation damping of the continuum medium. This step is used to guarantee the scattered waves can pass through the boundary without wave reflection. The way of developing artificial boundary can be divided into two categories in general: accurate and local. The first one requires the artificial boundary to meet the field equation of infinite medium, physical boundary condition, and radiation condition. This kind of boundary is accurate in finite element method, and it can also be set on the boundary of irregular structures and surrounding medium. In most cases, it is quite effective; however, the kinematic coupling of all the nodes on the artificial boundary can lead to the tremendous calculation storage and computation time. Compared to

5.6 Viscoelastic Artificial Boundary

the accurate boundary, the local boundary is not strictly accurate in theory, but it can also meet the accuracy requirement as long as the applied local boundary is suitable. The prominent characteristic of the local boundary is the coupling solutions require small calculation storage and short computation time, which is quite economical.

Currently, there are several popular artificial boundaries: viscous boundary, consistent boundary, superposition boundary, paraxial boundary, transmitting boundary, dynamic mapping infinite element, and so on. Among these boundaries, the viscous boundary, paraxial boundary, and transmitting boundary belong to the local artificial boundary in time domain. The paraxial boundary and transmitting boundary are more accurate. In the practical application, the high-order formula is complicated, so the accuracy of these two boundaries is limited to the second order. Though the damping boundary only has first-order accuracy, it is concise and more convenient for coding execution. Therefore, it is widely applied as well. The shortage of damping boundary is that it can only absorb the energy of scattered waves, while it cannot simulate the elastic restoring force of the semi-infinite medium. From the physical perspective, the mechanical model with the viscous boundary is actually a floating body. The low-frequency load can cause the drifting of the whole model. In the model of this chapter, viscoelastic artificial boundary is applied for it can concomitantly simulate the radiation of scattered waves and elastic restoring force.

In the practical problems, the scattered waves are diffusing in a geometric way, so the assumption of cylindrical waves is more reasonable. We adopted cylindrical coordinates, if the medium is cutoff at the radium R_b, we can arrange corresponding physical components (viscous damper C_b and the linear spring K_b) on the interceptive boundary:

$$C_b = \rho V_s \quad K_b = 0.5G/R_b \tag{5.53}$$

where ρ, G denotes the density and shear modulus of the medium, R_b denotes the distance between the boundary to the wave source, and V_s denotes the shear wave velocity.

If R_b can be determined accurately, the parameters related to the physical components of the artificial boundary can be obtained through formula (5.53). According to the parameters, we can utterly eliminate the reflection of the scattered waves, which means the model can simulate the propagation of the vibration wave from the finite field to the infinite field (Liu and Lv 2000). The viscoelastic artificial boundary consists of viscous damper and spring; if $K_b = 0$, it turns into Lysmer viscous boundary. In the finite element analysis for the interaction of soil and structure, the straight artificial boundary is more convenient. In the development of artificial boundary, the viscous damper is irrelevant to the distance between source and boundary (R_b), while it is relevant to the spring stiffness K_b. In this model, considering the attenuation of the dynamic stress with the depth and the convenience of programming, the distance R_b in different directions takes the same value. Due to the symmetrical feature, the right boundary is defined as the symmetrical boundary.

5.7 The Time Integral Step

Transient dynamic analysis (time history analysis) is applied to study the dynamic response of the numerical model under random time and load. Through this method, the time history of the displacement, stress, and strain under dynamic load can be figured out. The basic equation of motion for the transient dynamic analysis is given:

$$[M]\{\ddot{u}\} + [C]\{\dot{u}\} + [K]\{u\} = \{F(t)\} \tag{5.54}$$

where

$[M]$ = mass matrix
$[C]$ = damping matrix
$[K]$ = stiffness matrix
$\{\ddot{u}\}$ = node acceleration vector
$\{\dot{u}\}$ = node velocity vector
$\{u\}$ = node displacement vector

For an arbitrary time t, this equation can be considered as the statics equilibrium equation which involves the inertia force and damping force. In the finite element analysis, these equations are solved on discrete time points by Newmark integral method. The time increment between two adjacent time points is called integral time step. The accuracy of the transient dynamic analysis is determined by the length of the integral time step: the smaller the integral step is, the higher accuracy the method has. However, too small integral time step will occupy massive computer resource. Several criteria are referred in determining the right integral time step:

1. The integral time step should be small enough to solve the dynamic response of the model.
2. The integral time step should be small enough to follow the load function for obtaining the load-time relation.
3. The integral time step should be small enough to capture the waves when they are propagating among elements.

5.8 The Results of Transient Analysis

In the analysis, effective stress is applied, and the typical waves of the soil's dynamic response are demonstrated in Fig. 5.13. The numerical results are consistent to the field test in Chap. 2 in general, which proves the numerical results are reliable. However, the uncertainty in the real subway system creates some discrepancies between the field test data and the simulated results. Compared with the amplitudes of stress waves in Fig. 5.13, the amplitudes obtained from field test (Fig. 2.5) are different. Because these waves are corresponding to wheel sets, the different

5.8 The Results of Transient Analysis

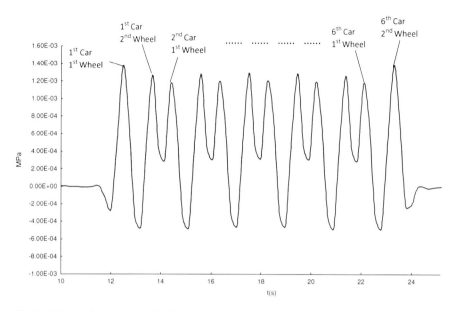

Fig. 5.13 Typical wave curve of soil response obtained through numerical simulation

weights of each carriage can cause different wheel-rail interaction forces which can eventually lead to the different amplitudes of the stress waves. The response curves of soil in numerical model are smoother than the curves obtained in the field test. This phenomenon is caused by the inhomogeneity of tunnels and soils in field.

We chose the cross section in numerical model which is corresponding to the field test; when the train is passing this section, the response contour plot of horizontal stress σ'_x is demonstrated in Fig. 5.14. It can be figured out that the distribution of σ'_x is concentrated around the tunnel. For illustrating this problem further, the data from the field test and the numerical model is compared in Table 5.5.

According to Tables 5.5 and 5.6, we found that the response stress amplitude in numerical model is generally larger than the field test. Two reasons can explain this phenomenon: firstly, the weight of the train applied in the numerical model is considered as the maximum value in real cases; secondly, the real subway system constantly adopted vibration absorption and reinforcement measures which can prevent the propagation of the dynamic load. Actually, the numerical model is calculated under the most unfavorable conditions. Therefore, this model can be used in evaluating the influence range of subway vibration.

The contour plot of response amplitude of horizontal stress σ'_z in longitudinal section is shown in Fig. 5.15. According to the calculated results, σ'_z propagates further in vertical direction, while σ'_x propagates further in horizontal direction. For determining the influence range of the dynamic load of train, we set the threshold

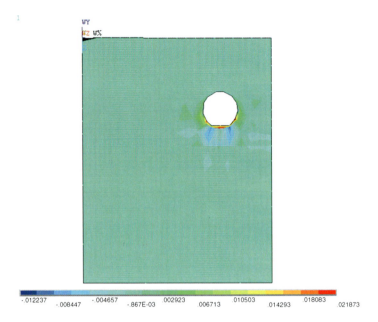

Fig. 5.14 Contour plot of response amplitude of horizontal stress in cross section

Table 5.5 Comparison of response amplitude of stress between calculation and field test along the depth at 1.8 m from the tunnel edge

Depth (m)	Response amplitude of σ_x' in numerical model (kPa)	Response amplitude of σ_x' in field test (kPa)
8.5	0.23	0.16
11.5	0.61	0.48
13.5	1.13	0.79

Table 5.6 Comparison of response amplitude of stress between calculation and field test along the horizon at 13.5 m depth

Horizontal distance (m)	Response amplitude of σ_x' in numerical model (kPa)	Response amplitude of σ_x' in field test (kPa)
1.8	1.13	0.79
3.8	0.82	0.69
6.8	0.64	0.45

value of the dynamic stress is 0.01 kPa. If the stress is under the threshold value, we consider the influence of the dynamic load on this area as ignorable. In the horizontal direction, the horizontal stress σ_x' can affect the area which is 12.78 m from the lateral rim of the tunnel; in the vertical direction, the vertical stress σ_z' can affect the area which is 18.17 m from the bottom of the tunnel.

5.9 Chapter Summary

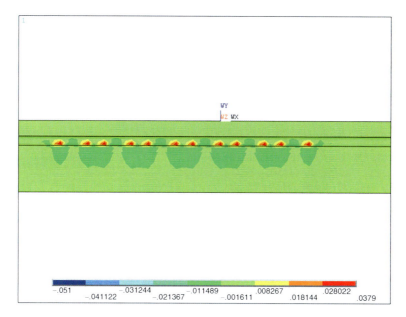

Fig. 5.15 Contour plot of response amplitude of horizontal stress in longitudinal section

5.9 Chapter Summary

This chapter primarily conducted the finite element analysis of the dynamic response of soil around the tunnel under dynamic load induced by the subway train; the main work and conclusions are summarized as below:

1. This chapter analyzed the mechanism of the generation of dynamic load induced by subway train and simulated it by programming in ANSYS.
2. According to the field investigation and data from the boreholes, we developed a 3D finite element numerical model by inputting the relevant data.
3. For obtaining accurate results without too large computational domain, artificial boundary was applied to simulate the soil body in the discretization of this model.
4. The dynamic response of the soil in the model was consistent to the data obtained from the field test. And the results were proved effective and reliable.
5. According to the calculation results, σ_z' propagates further in vertical direction, while σ_x' propagates further in the horizontal direction. Furthermore, the influence range of the dynamic load was determined by this model.

References

Andersen L, Jones CC (2006) Coupled boundary and finite element analysis of vibration from railway tunnels—a comparison of two and three-dimensional models. J Sound Vib 293(3):611–625

Balendra T, Koh CG, Ho YC (1991) Dynamic response of buildings due to trains in underground tunnel. Earthq Eng Struct Dyn 20:275–291 (in Chinese)

Cheng B (2003) Study on the settlement of shield tunnels due to metro train's vibrating loads. PhD dissertation, Tongji University (In Chinese)

Feng JH, Yan WM (2008) Numerical simulation of the random vibration load of the train. J Vib Shock 27(2):49–52 (in Chinese)

Feng JH, Yan WM, Chen XM (2007) Contrastive analysis of the different load simulation methods. Railw Eng 10:79–81 (in Chinese)

Gao F (1998) Analysis of dynamic responses of a railway tunnel subjected to train loading. J Lanzhou Railw Inst 17(2):6–12 (in Chinese)

Huang JH, He CH, Yang GT, Liu WD (1999) FEM analysis of the vibration wave in the foundation. J Vib Shock 18(1):38–43 (in Chinese)

Li DW (1998) A deterministic analysis of dynamic train loading. J Gansu Sci 10(2):25–29 (in Chinese)

Liang B, Cai Y (1999) Dynamic analysis on subgrade of high speed railways in geometric irregular condition. J China Railw Soc 21(2):84–88 (in Chinese)

Liu JB, Lv YD (2000) Study on an analysis method of dynamic soil-structure interaction based on interface idea. In: Chen ZP, Jiang JJ (eds) Research reports of structural engineering and vibration, vol 5. Tsinghua University Press, Beijing, pp 1–17 (In Chinese)

Luo YY, Gen CZ (1999) Influence of track state on vertical wheel/track dynamic overloads. J China Railw Soc 21(2):42–45 (in Chinese)

Pan CS, Xie ZG (1990) Measurement and analysis of vibrations caused by passing trains in subway running tunnel. China Civil Eng J 23(2):21–28 (in Chinese)

Pan CS, Li DW, Xie ZG (1995) The study of the environmental effect of the vibration induced by Beijing subway. J Vib Shock 14(4):29–34 (in Chinese)

Saeed M (2008) Finite element analysis-theory and application with ANSYS, 3rd edn. Upper Saddle River, Prentice Hall

Wang XC, Shao M (1997) Basic principle and numerical method of finite element method, 2nd edn. Tsinghua University Press, Beijing

Wang FC, Xia H, Zhang HR (1999) Vibration effects of subway trains on surrounding buildings. J Beijing Univ Technol 23(5):45–48 (in Chinese)

Wang XL, Yang LD, Gao WH (2005) In-situ vibration measurement and load simulation of the raising speed train in railway tunnel. J Vib Shock 24(3):99–102 (in Chinese)

Xia H, Zhang N, Cao YM (2005) Experimental study of train-induced vibrations of environments and buildings. J Sound Vib 23(3):1017–1029

Xie WP, Sun HG (2003) FEM analysis on wave propagation in soils induced by high speed train loads. Chin J Rock Mech Eng 22(7):1180–1184 (in Chinese)

Xu ZY, Shen ZJ (1981) 2-D dynamic analysis of effective stresses of seismic liquefaction. J East China Collage Hydraul Eng 3:1–14 (In Chinese)

Yan WM, Zhang W, Ren M, Feng J, Nie H, Chen JQ (2006) In situ experiment and analysis of environmental vibration induced by urban subway transit. J Beijing Univ Technol 32(2):149–153 (in Chinese)

Zhang YE, Bai BH (2000) The simulation of the vibration load on the tunnel induced by subway train. J Vib Shock 19(3):68–78 (in Chinese)

Zhang YE, Pan CS (1993) Tests and analysis of the dynamic response of tunnels subjected to passing train load. J Shijiazhuang Railw Inst 6(2):7–14

Chapter 6
Settlement Prediction of Soils Surrounding Subway Tunnel

6.1 Introduction

Sufficiently utilizing the underground space resources in urban areas, subway system has become an important urban infrastructure due to its advantages of convenience, safety, high-speed. However, the construction and operation of the subway system are quite influential on subway structure and surrounding environment which can lead to many problems. Among these problems, the deformation of surrounding soils has been highlighted because it can eventually cause the settlement of the subway tunnel and ground. The deformation of surrounding soil could be divided into two stages: (1) deformation and settlement emerged during the process of tunnel construction and (2) deformation and settlement caused by subway dynamic loading during operation stage. Taking Shanghai subway as an example, because of the wide distribution of soft clay in Shanghai area, ground settlement is inevitable during the tunnel excavation regardless of the tunneling methods. Meanwhile, some monitoring results (Chen and Zhan 2000) showed large axial differential settlement occurred in Shanghai Metro Line 1 during its operation. Internationally, researchers (Prevost 1978; Zienkiewicz et al. 1985; Dafalias 1986; Dafalias and Herrmann 1986; Miura et al. 1995; Li and Selig 1996; Chai and Miura 2002; Sakai et al. 2003; Indraratna et al. 2009; Tang et al. 2011) have already conducted many studies on residual deformation of soils under traffic loads and proposed some models, but there is still no model that is fully approved and accepted because these models are either too complex to apply in practices or too simple to rightly reflect the accumulated residual deformation of soils. Therefore, it is a valuable work to establish an appropriate prediction model which can use the existing monitoring data to predict settlement for a period of time in the future.

This chapter is organized as follows: Sect. 6.2 analyzed the influence of tunnel excavation on ground settlement. In Sect. 6.3, factors affecting long-term settlement of subway tunnel in soft soil area and its mechanism were studied. Section 6.4 proposed a non-isochronous nonhomogeneous exponential gray model

(NNGM) (1, 1) to predict cumulative plastic deformation of soils and axil settlement of subway. Effective stress analysis for long-term settlement of the subway tunnel was discussed in Sect. 6.5, and finally, conclusions were drawn in Sect. 6.6.

6.2 Analysis on Subway Tunnel Settlement During Tunneling in Soft Soil

6.2.1 Factors

The tunnels of Shanghai Metro Line 1 are primarily located in mucky clay layer. The soil in this layer is characterized with high sensitivity; therefore, once the soil is disturbed, the soil strength will be decreased dramatically and the soil will reconsolidate; it will take a long time to reach the equilibrium state. Generally, the ground settlement caused by tunneling in this layer can be divided into three stages (shown in Fig. 6.1).

The first stage is caused by various construction factors in the process of excavation. In this stage, the strata deformation can be considered generated under undrained conditions. The second stage is the primary consolidation settlement caused by the dissipation of excess pore water pressure created in the first stage. The third stage is the secondary consolidation caused by the creep of soil skeleton. This consolidation starts after the excess pore water pressure is fully dissipated. Generally, the first stage settlement is called pre-settlement or construction settlement, and the second and third stages are called post-settlement.

For the pre-settlement caused by tunneling, it can be divided into three stages (Fig. 6.1):

1. Soil in front of the shield tunneling machine (STM) has obvious forward and upward movements because of the extrusion, which eventually leads to the uplift of surface soil.
2. The ground surface above the STM sinks slightly when the shield advances.
3. Settlement could be induced by the structural space between the outer wall of shield tunnel segments and soil body after the STM passed.

The pre-settlement generally completed within 1–2 months after the construction finished. Its influence factors mainly originate from the following three aspects:

1. Heterogeneous distribution of the underlying stratum
   Heterogeneous distribution of the underlying stratum is the basic reason of uneven vertical deformation of tunnels. In practical engineering, properties of the soil along the tunnel axis are different. Meanwhile, the layering and transition of the soil and the buried depth of tunnel are always in change. Because of the different properties of the soil along the tunnel axis, the disturbance degrees, resilience values, primary consolidation and secondary consolidation on settlement, settlement rate, and settlement time are all different, which lead to the uneven settlement along the tunnel (Sun 2006; Liao and Hou 1998). In general,

6.2 Analysis on Subway Tunnel Settlement During Tunneling in Soft Soil

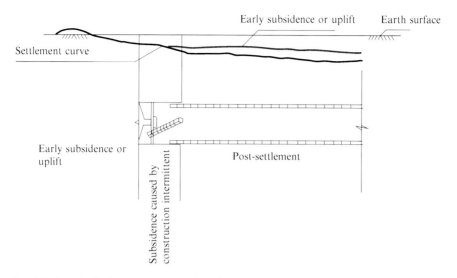

Fig. 6.1 Longitudinal settlement curve along the tunnel axis

the place where the properties of the stratum change the most is the place where the largest uneven settlement of the tunnel occurs.

2. Loading and unloading processes on the ground surface above the tunnel
   Surface loading above the subway tunnel could produce uneven settlement for the tunnel located in the soft soil layer. Especially, this kind of uneven settlement will continuously increase under the additional stress generated by a large area of load applied to a thick and soft soil layer. The compression index of soils beneath the tunnel decreases after tunnel constructions because the self-weight stress of overlying soils reduces during tunnel excavation. Moreover, due to the construction disturbance surrounding the tunnel, the long-term secondary consolidation of the underlying soils still continues when the surface loading is applied. It indicates that it is necessary to consider the effect of a large area loading above the tunnel on the uneven settlement along the tunnel in order to ensure the safety of normal operation of subway.
3. Construction load adjacent to the tunnel
   (a) Building load near the tunnel
       Subway tunnel generally goes through the downtown areas where high-rise buildings are constructed densely. These high-density and high-rise buildings, especially for those distributed along the subway, could generate large additional stress which has significant effect on settlement. Moreover, the properties of the soil beneath the tunnel vary along the tunnel axis and so do the thickness of the compressible stratum, which leads to significantly different reactions of soil consolidation under different additional stresses. These different reactions could produce longitudinal uneven settlement of the tunnel.

(b) Foundation excavation near the tunnel

High-rise building basements usually adopt the deep foundation excavation method which is actually an unloading process. Generally, deep foundation excavation mainly has two impacts on the tunnel: Firstly, the lateral displacement of retaining structures and uplift in the bottom caused by excavation could induce the soil settlement outside the foundation and then result in tunnel settlement. Secondly, flexural deflection of the subway tunnel will occur due to lateral horizontal displacement of retaining structure toward the inside of the foundation pit. Therefore, the tunnels which have different distances to the foundation pit will demonstrate significant longitudinal differential settlement.

(c) Subway shield tunneling near the existing tunnels

Because of the requirement of three-dimensional integrated underground space development and efficient subway system, more tunnels are constructed closely to the adjacent tunnels. The newly built tunnels will produce a settling tank which has approximate normal distribution in surrounding soil due to its disturbance and result in longitudinal uneven settlement of the existing tunnel.

Relevant data shows that the large subway tunnel settlement tends to appear in the area where high-rise and high-density buildings are located and constructed. It is necessary to control the various construction activities near the tunnel and monitor the tunnel strictly for ensuring the tunnel safety and subway vehicles operation. Therefore, it is stipulated in Shanghai Metro Protection Technical Standard that total displacement of the metro tunnel induced by loading and unloading of the surrounding environment shall not exceed 20 mm.

6.2.2 Scope of Settlement and Settlement Tank

Tunnel excavations not only cause settlement above the tunnel axis but also cause some settlement in the surrounding soil. In theory, ground settlement of the cross section along the tunnel axis fits to the Gauss normal distribution (Fig. 6.2).

However, the settlement is controlled by many factors with complex relationship among them in practice. Therefore, the settlement curve may not meet a certain function. Through measured settlement data of some observation points on cross section along the tunnel axis, we can draw out the development of settling tank with time (Fig. 6.3).

Through analyzing of the three settling tanks above, it can be concluded that:

1. In the front of excavation faces, some settling tank is convex because of ground upheaving, such as settling tank curves on cross sections A3, A9, and A15 (note: the distances from cross sections A3, A9, and A15 to the end shaft are 14.00 m, 38.00 m, and 61.00 m, respectively). After the STM passing, ground sinks gradually and the settling tank turns into concave.

6.2 Analysis on Subway Tunnel Settlement During Tunneling in Soft Soil

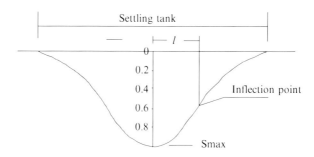

Fig. 6.2 Gaussian normal distribution of settling tank

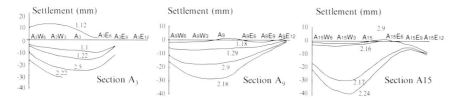

Fig. 6.3 Filed measured curve of settling tank

2. Generally, centers of these settling tanks overlap with the centers of the tunnel, but in some cases, there are deviations, such as the measured settlement curves of cross section A3. The reason may be because the soil on the top of tunnel was suffering more disturbance than the surrounding under tunneling pressure. When the surrounding soils move toward the construction gap, the transfer of displacement in disturbed soil reduces greatly due to the plastic failure.
3. The influence scope of settling tank is about 15 m on both sides of the tunnel, and the main settlement area lies within 9 m of both sides. Over this range, the settlement is generally less than 10 mm and does not have too much impact on the ground.
4. The angle between the slope of the settlement curve and the horizontal line is about 50°, equal to about $45° + \psi/2$, where ψ is the internal friction angle of mucky clay layer which the tunnel passes through, and its value usually equals to 9.6°. It indicates that the settlement curve is similar to the sliding surface under active earth pressure.

6.2.3 Relationship Between Settlement and Grouting Quantity

As previously mentioned, timely grouting the construction gap has an impact on ground settlement, and the real construction experience also proves this point. Although the ground settlement is controlled by various factors, the grouting

Fig. 6.4 Theoretical curve of grouting filling rate and land subsidence

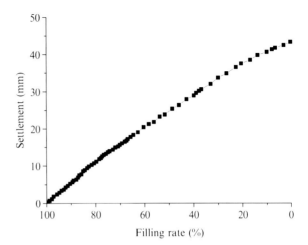

quantity is crucial. Ground settlement can be controlled effectively by controlling the grouting quantity.

Taking the tunnel excavation between Shanghai Stadium Station and Xujiahui Station of Shanghai Metro Line 1, for example, the outer diameter of shield tunneling machine is 6.34 m, and the outer diameter, inner diameter, and width of retaining structure are 6.20 m, 5.50 m, and 1.00 m, respectively. So the volume of the construction gap in unit width is

$$V = \left(\frac{\pi}{4} \times 6.34^2 - \frac{\pi}{4} \times 6.20^2\right) \times 1 = 1.40 \text{m}^3$$

If this construction gap isn't grouted, the volume of settling tank is equal to the ground loss. Based on the analysis of the curve of settlement, some control parameters could be obtained.

Ground loss in unit width: $V = 1.40 \text{ m}^3/\text{m}$
The width parameters of settling tank: $i = R \cdot \left(\frac{Z}{2R}\right)^{0.8} = 3.10 \times \left(\frac{12.00}{2 \times 3.10}\right)^{0.8} = 5.26\text{m}$
The maximum settlement: $S_{max} = \frac{V}{2.50 \times i} = \frac{1.40 \times 1000}{2.5 \times 5.26} = 106 \text{ mm}$

Because construction gap is grouted, the settlement decreases significantly. Generally, filling rate is used to characterize the grouting quantity. The filling rate is equal to the ratio of grouting quantity and the volume of construction gap.

Theoretical curve of grouting filling rate and ground settlement is showed in Fig. 6.4.

But in fact, filling rate is all higher than 100 %, and it is even common to be more than 200 % during the construction of Shanghai Metro Line 1. The relationship between measured settlement along the axis of tunnel and grouting quantity is showed in Table 6.1. The separated water volume of the slurry in consolidated process takes up about 6.3 % of the slurry volume as the slurry consistency is 9–11.

6.2 Analysis on Subway Tunnel Settlement During Tunneling in Soft Soil

Table 6.1 Relationship between measured settlement and grouting quantity along the axis

Ring number	Grouting quantity (m³)	Filling rate (%)	Settlement (mm)
16	3.0 × 93.7 %	201	−33.33
20	3.3 × 93.7 %	221	−33.32
24	4.0 × 93.7 %	268	−28.16

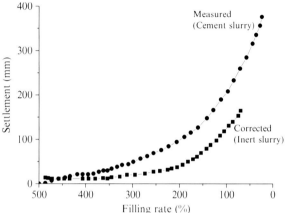

Fig. 6.5 Curves of grouting filling rate and measured settlement

This kind of water seeps into surrounding layers. So the volume of practical filling slurry is equal to the practical grouting quantity multiply by (100−6.3 %). (Note: the data 6.3 % was obtained from laboratory experiment.)

Because of the limited data, the subsidence cannot be obtained under different filling rates. But we can get two results at least:

1. The settlement decreases with grouting quantity increase.
2. When grouting quantity is more than 200 %, the settlement does not change with grouting quantity obviously. Therefore, it's not necessary to improve grouting quantity aimlessly which is proved waste of material.

In the practical constructions, we control the filling rate of grouting quantity within 200 %, that is, less than 2.80 m³, and then adjust other shield parameters such as soil quantity and advancing speed to control settlement within theprescribed scope (+10 mm to −30 mm).

Tang et al. (1993) studied the relationship between the tunnel axis heave and grouting quantity and concluded that it is best to control grouting quantity at the level of about 2.50 m³. According to the two aspects of the results presented above, the grouting amount should be 2.5–2.8 m³ when slurry consistency is between 9 and 11. In the grouting experiments of Shanghai Metro, we got the relationship between settlement and filling rate. In these experiments, the grouting slurry we used is cement, while the inert slurry is usually taken in the grouting in the practical construction. These two kinds of grouting have different impacts on settlement. Then we use practical construction data to correct it which is showed in Fig. 6.5.

Table 6.2 Comparison between calculated value and measured value

Point	X (m)	Grouting (m³)	g	Settlement measured (mm)	Calculated value (mm)
A_3	0	3.0	0.045	−31.33	−31.64
A_3W_3	3			−26.32	−26.76
A_3W_6	6			−16.35	−16.20
A_6	0	3.3	0.043	−30.32	−30.27
A_6W_3	3			−26.12	−25.81
A_6W_6	6			−15.92	−15.50
A_5	0	4.0	0.04	−28.16	−28.20
A_5W_3	3			−18.84	−23.86
A_5W_6	6			−9.30	−14.46

6.2.4 Estimation of the Ground Pre-settlement

Known from the analysis of the previous study, there're many factors to influence the ground settlement. And these factors are related to each other and vary obviously, so it is difficult to propose a precise mathematical model. However, on the cross section along the axis of the tunnel, the settlement distributes in trough type. As a result, an estimation model could be established by considering the following aspects:

1. Tunnel geometries, such as depth and diameter
2. The estimation points of subsidence
3. Factors of construction

The first two are easy to determine, but it's difficult for the last one which is controlled by many factors. In actual estimation, we often use stratigraphic gap coefficient g to evaluate the ground loss. The g reflects and contains all kinds of construction factors. So we should choose g on the basis of different soil properties and construction conditions. For example, the gap factor should be increased if overexcavation is generated by shield correction. And we can also control grouting by decreasing gap factor. Based on previous experiences combined with the measured data of the tunnel excavation between Shanghai Stadium and Xujiahui Station of Line 1, an estimated model is proposed:

$$\delta = \frac{0.627 \cdot D \cdot g}{H \cdot (0.956 - H/24 + 0.3g)} \cdot \exp\left(\frac{-6 \cdot X^2}{30 \cdot (6 - 5/H) \cdot (2 - g)}\right) \quad (6.1)$$

where δ is the pre-settlement, D is diameter of tunnel, H is the tunnel depth, g is stratigraphic gap coefficient, and X is the distance to tunnel axis.

Comparison between calculated value and measured value is showed in Table 6.2 (taking the total 9 measured points of three cross sections A_3, A_9, and A_{15} along the tunnel axis in consideration).

6.2 Analysis on Subway Tunnel Settlement During Tunneling in Soft Soil   199

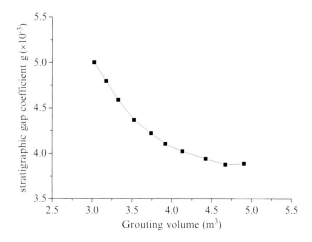

Fig. 6.6 Curve of gap coefficient and grouting volume in practice

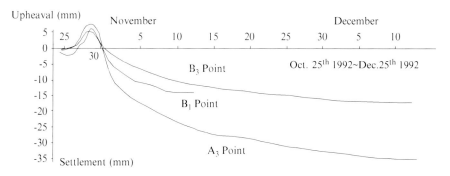

Fig. 6.7 The settlement curve at B5 monitoring point of No. 2 shield (lasting 50 days)

It follows that calculated values match the measured ones very well. Although some deviation happed in the cross section A5, it is because soil disturbed on the cross section did not finish reconsolidation. In the proposed model, g is a key parameter; it has a large number of control factors while limited information is available. Hereby, the only way to determine it is on a basis of grouting volume. From Table 6.2, we can know g decreases with grouting increase. The relationship is shown in Fig. 6.6.

In Fig. 6.7, the point of B_3 is a settlement curve plotted in the case of grouting volume 2.60 m^3 and lasts 50 days. Its maximum settlement is 19 mm. It is the best scheme determined after synthetically considering all kinds of factors that cause ground settlement when shield advances in the soft soil; it can control ground settlement successfully (shown in Fig. 6.7).

6.2.5 Summary of Early Settlement's Analysis

Based on the above five parts analysis, some important conclusions could be drawn as follows:

1. In soft soil area, ground settlement caused by tunnel excavation is controlled by many factors such as grouting, volume of excavated earth, the soil pressure, and advance speed. To make influence on the surrounding environment under control, it is necessary to take all of the factors into consideration to make appropriate construction projects.
2. Grouting volume and slurry consistency are crucial factors to control ground settlement. Considering the heave effect on the tunnel axis which grouting treatment has, the volume of grouting should be between 2.5 and 2.8 m^3 when the slurry consistency is between 9 and 11.
3. Besides the effect of the stratum loss and construction gap on ground settlement, we should also consider the deformation of the tunnel segment and other factors.

6.3 Analysis of Long-Term Settlement of Subway Tunnel in Soft Soil

6.3.1 Factors Affecting Long-Term Settlement of Subway Tunnel in Soft Soil

Settlement of subway tunnel is affected by various factors. Tunnels which are located in the saturated soft soil generally increase longitudinal settlement continuously in long-term operation, which is likely to account for the main part of the total settlement. For example, in the Shanghai Metro Line 1, completed and put into operation in 1995, the settlement monitoring finds that long-term tunnel settlement in the operation and the uneven settlement are quite large; moreover, they are still developing now, and there is no convergence trend (Chen and Zhan 2000; Ren et al. 2012). Preliminary analysis shows that the main settlement influence factors of the subway tunnel in soft soil include the following several aspects:

1. Vibrations of subway traffic
   During normal operation, subway tunnels undergo long-term cycles of train vibration load. Study shows that the structural vibration displacement caused by the train vibration load is quite small; bending, axial force, and shear force does not exceed 10 % of the corresponding value caused by the water and soil pressure. But under the effect of train vibration over a long period of time, attention should be paid to the possibility of liquefaction of saturated sandy soil layer and the vibrated subsidence of saturated clay under the tunnel (Wang et al. 2001).

2. Differential settlement of subway tunnel and the stations

   Subway tunnel and station differ greatly in the structure performance, geological environment, construction method, operating conditions, and other aspects. Thus, there inevitably exists differential settlement between the tunnel and station. Sometimes the differential settlement is even large enough to cause the leaking of tunnel girths and cracking of segments. Subway station is reinforced concrete structure of long strip box, and the station often built in underground two to three stories, so soils at the bottom of the station is in severe unloading condition. Subway station is usually constructed by opencut method in the underground diaphragm wall or inverse construction method. The structure body of the station is connected entirely with the underground diaphragm wall. The soil deflection of the construction is small. And the subway tunnel construction leads to unloading at the bottom of the tunnel, but it is much less on the quantity. The construction disturbance on the surrounding soft soil leads to downward settlement of the tunnel; it results in uneven settlement between the station and tunnel (Tong 2000).

3. Comprehensive effect of land subsidence in the city

   Tunnels in the soil are inevitably affected by urban land subsidence. Urban land subsidence is a worldwide problem to be solved. Especially for the large- and medium-sized cities in China, land subsidence problems are very serious. Monitoring data show that in the central area of Shanghai city, from 1990 to 1998, the average cumulative land subsidence is 135 mm, with an annual average of 15 mm; local area is even larger. Due to the stratigraphic structure in soft soil area, subsidence funnels appeared frequently. When tunnels are constructed across subsidence funnel area, the settlement of the tunnel located in the funnel area is obviously larger; longitudinal uneven deformation can be more serious in the long term.

4. Earthquake

   Due to the soil-structure interaction, the mechanism of earthquake and structure seismic response is quite complicated. Due to the geological environment difference, the tunnel axis deflection, effect of adjacent constructions, the different surrounding constraints, etc., the dynamic responses to earthquake vary for different tunnel sections. This fact can lead to severe longitudinal uneven deformation of the tunnel (Yu et al. 1986).

For tunnels in soft soil, more attention should be paid to the liquefaction of saturated silt and fine sand in earthquake. The influences of liquefaction on the longitudinal uneven deformation of tunnels are mainly manifested in massive soil seismic subsidence or upward buoyancy generated by liquefaction. These will result in huge uneven vertical tunnel structure deformation and increase the structure internal force of the tunnel rapidly and eventually lead to the failure of tunnel structure. Thus, it is necessary to take some appropriate measures to improve the weak geological conditions and strengthen the tunnel structure in the stratums where liquefaction is prone to occur (Dai 2008).

Zheng et al. (2005) have calculated the seismic subsidence of Shanghai Subway in soft soil; the calculation results show that the tunnel settlement is less than 5 mm. Wang et al. (2006) calculated and analyzed the influence of deep excavation on adjacent subway tunnel; the maximum lateral displacement is about 13 mm, which is in the allowable range for the subway protection. Zhou and Wang (2002) applied mechanical and seepage coupling method, from the aspect of tunnel leakage, to calculate the tunnel subsidence of soft soil surrounding the tunnel under different permeability-coefficient conditions; they analyzed the stability time for the settlement caused by the leakage and showed that subsidence settlement caused by tunnel leakage can reach more than 200 mm, which is approximate to the settlement of subway tunnel constructed in the soft area of Shanghai.

6.3.2 Mechanism Analysis of Long-Term Additional Settlement in Foundation Caused by Subway Load

Previous studies show that (Liu and Hou 1991) saturated soft clay possesses the threshold dynamic stress ratio and the critical dynamic stress ratio. With increase in the number of cyclic loading, soil only produces elastic deformation when the dynamic stress ratio is less than the threshold ratio. When the dynamic stress ratio is larger than the threshold ratio, both elastic and plastic deformation will occur, the increment rate of deformation will decrease, and deformation tends to be a stable value at last. While when the dynamic stress ratio is over the critical stress ratio, the deformation augments gradually with the cycle's number and finally gets into failure. The similar results are obtained by dynamic triaxial tests presented in this book.

The previous analysis showed that the dynamic stress under traffic loads decreased along the soil depth. It indicates there is a critical depth where dynamic stress ratio is equal to the ground soil threshold dynamic stress ratio. Below this critical depth, dynamic stress ratio is less than the threshold of soil dynamic stress ratio, and the induced deformation of soils cannot be accumulated regardless of loading time. Therefore, we just need to consider the deformation of soils within this critical depth when we analyze the long-term additional settlement of the tunnel.

From the test we can know that the soil within the critical depth under the cyclic load will be softening. However, it occurs only in a part of the soil at first. Then, because of the strength loss of the softened soil, the additional stress is undertaken by surrounding area where soil has not been softened yet. Under a period of cyclic loading, the surrounding area is softened like the initial softened soil. The area of softening soil gradually expands and eventually reaches the critical boundary in soil. In this process, accumulative plastic deformation and soil structure destruction result in massive deformation. This is the reason that the additional settlement of soil is large and grows in the long term under dynamic train load.

In practical engineering, to avoid the long-term additional settlement of soil, the ground treatments to the soil within critical depth are necessary. Generally, the fracturing grouting reinforcement is a more mature and convenient method. Firstly, it can change the properties of soil beneath the subway tunnel. Secondly, it also can change the stress distribution in foundation soil and narrow down the critical boundaries. Thirdly, fracturing grouting reinforcement can also play an important role of compensating the settlement. Finally, the most important advantage is that it does not affect the normal operation of the subway. Based on the above analysis, the fracturing grouting treatment has achieved a great effect on the long-term settlement control for the soil foundation within 5 m below the tunnel.

6.4 Applications of Gray Prediction Theory in Predicting Long-Term Settlement of Subway Tunnel

The gray system theory is firstly proposed by J.L. Deng in 1982, and one of its most successful aspects is its application in prediction in greatly wide fields. Since then, the theory has become quite popular with its good performance to deal with the systems that have partially unknown parameters (Kayacan et al. 2010). Compared with conventional statistical prediction models, one of the major advantages of gray model is that it requires only a limited amount of data to predict the system behavior (Deng 1989). In gray prediction theory, by virtue of accumulated generating operation (AGO), smooth discrete data can be transformed into a sequence having the approximate exponential regularity, which is the essence of the prediction model. During the last two decades, the gray system theory has been widely and successfully applied to various scientific areas such as social and economic, agricultural, engineering, military, and transportation. A large number of empirical applications (Liu et al. 2010) indicate that the gray model can achieve high prediction accuracy for exponential and approximate exponential systems.

6.4.1 Research Backgrounds and Introduction

Unrecoverable permanent deformation could be generated as a form of mutual sliding which occurs between the soil particles; and new arrangement of them is formed under cyclic loads such as subway, high-speed railway, and operation of heavy machinery. This deformation can be accumulated over time and could bring land subsidence, and seriously affects the normal use of the structures subjected to the cyclic loads, or even destroys them. For example, Line 1, Shanghai Metro was built in 1995. The tunnel settlement had reached up to 155 mm in just 4 years and induced some corresponding land subsidence, shown in Fig. 6.8. Miura et al. (1995)

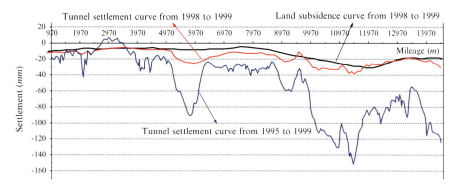

Fig. 6.8 Accumulated settlement of Line 1, Shanghai Metro from 1995 to 1999 (After Chen and Zhan 2000)

found that the additional monitoring settlement of Saga Airport road reached about 150 mm in 2 years.

Knowledge of cumulative plastic deformation and failure mechanisms for soils under cyclic loading is essential for the proper design and maintenance planning of highway pavements, railway tracks, airport roads, offshore constructions, etc. Being able to predict cumulative plastic deformation of soil subjected to cyclic loads has been quite a popular and significant subject for researchers, and a large number of experimental and theoretical studies have been conducted. Among them, two main methods can be summarized.

The first is numerical method. The process of this method is that constitutive models of soil are established through a certain approach firstly, such as bounding surface model (Dafalias 1986; Dafalias and Herrmann 1986), nested yield surface model (Prevost 1978), and elastic-plastic model based on the theory of generalized plasticity (Zienkiewicz et al. 1985). Then, the deformation characteristics of each cycle are simulated by some numerical methods on the base of established constitutive models. However, it has already been proved difficult to establish a constitutive model that could truly reflect the elasticity, viscosity, and plasticity of soils. And even with this model, the computation is so staggering that it has little possibility to be used in reality because these loads usually have thousands of cycles. The second method is empirical model. The essence of the empirical model is choosing suitable fitting function to simulate and predict it. Based on a large number of laboratory experiments or in-site tests, empirical prediction models of cumulative plastic deformation are built by fitting deformation and its factors such as consolidation ratio, over-consolidation ratio, dynamic stress amplitude, loading frequency, number of cycles, and cyclic stress ratio. This method is relatively easy to be applied in real-time projects. Thus, it attracts many researchers to study on it, and various models have been developed. Among them, one of the most commonly used is the following power model proposed by Monismith et al. (1975):

6.4 Applications of Gray Prediction Theory in Predicting Long-Term...

$$\varepsilon_p = A_0 N^b \qquad (6.2)$$

in which ε_p is cumulative plastic strain (%), N is the number of cycles, and A_0, b are two parameters depending on soil type, soil properties, and stress state.

Li and Selig (1996), Chai and Miura (2002), respectively, improved this model on the base of dynamic triaxial tests under different cases. In addition, some other empirical models or semiempirical models have been developed by a series of laboratory tests under different combinations of factors (Brown 1974; Hyde and Brown 1976; Hyodo et al. 1992; Sakai et al. 2003).

The above literature indicates that even though the existed researches provide important references for the understanding of cumulative plastic deformation; it still needs much study in this area. In practical applications, this case often happens: we need to take advantage of known data to predict the unknown. Another problem is how to choose a fitting function reasonably, especially when the simulated results of different fitting functions are similar for the same raw data of tests. It is not rigorous and reasonable if chosen at random. Our purpose is to try to present a simple theoretical prediction model with a wide range of applicability. This is aimed at developing a better understanding and insight into predicting cumulative plastic deformation of soils subjected to cyclic loads.

The essence of long-term settlement of subway tunnel is that the unrecoverable permanent deformation of foundation soils subjected to subway loading is generally accumulated over time. Therefore, a capable method to predict settlement of subway must be able to predict cumulative plastic deformation first. The period of cumulative plastic deforming of soils that are subjected to cyclic loads is very long, usually several years or even decades, especially for saturated soft clay. Thus, it is too difficult to get a large amount of data by continuously monitoring in such a long period of time. Although lots of data could be obtained by high-frequency monitoring within a short time, most of these data are nonrepresentative and incapable of building an effective prediction model because sampling time of data is too short compared with the duration of deforming. For these reasons, the capable deformation data actually obtained in real-time engineering is limited. The gray prediction model can just compensate for the lack of monitoring data. Moreover, the AGO could reduce the randomness of data and enhance the regularity of data series. A large number of cyclic loading tests (Liu et al. 2008; Li and Selig 1996; Indraratna et al. 2009; Wichtmann et al. 2010; Huang et al. 2006; Liu 2008) suggest that no matter what kind of soil is tested, its cumulative deformation follows an exponential regularity, as shown in Fig. 6.9. Coincidentally, using exponential function or approximate exponential function to fit and predict is the essence of the gray models. Therefore, it could be believed that the gray prediction model is a reasonable and potential approach, at least in theory, to simulate and predict cumulative plastic deformation of soils subjected to cyclic loads.

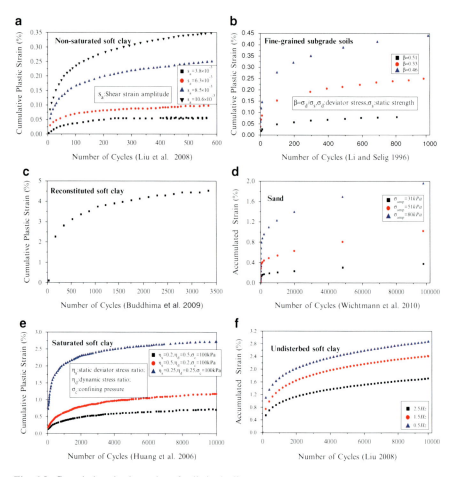

Fig. 6.9 Cumulative plastic strains of soils in the literature

6.4.2 Methodology of Gray Prediction Theory

In gray systems theory, GM (*n*, *m*) denotes a gray model, where *n* is the order of the differential equation and *m* is the number of variables (Liu et al. 2010). GM (1, 1) is the most widely used because of its simplicity and satisfactory accuracy. Several of the gray models were developed on the basis of GM (1, 1), such as GM (1, *n*) and discrete gray model DGM (1, 1). All of the above models require the 1-AGO data to obey the law of exponential growth with time sequence. However, Fig. 6.9 shows that the cumulative plastic strain of soils under cyclic loads follows a nonhomogeneous exponential distribution. The 1-AGO data also follows the nonhomogeneous exponential distribution but not exponential distribution. Thus, the above-mentioned gray models are not good enough to be directly used to predict the cumulative plastic deformation and long-term settlement subjected to cyclic loads.

6.4 Applications of Gray Prediction Theory in Predicting Long-Term...

The gray prediction model which requires the data used for modeling is isochronous. The obtained data from field monitoring, however, are usually non-isochronous data sequence. It is necessary to transform the non-isochronous data sequence into the available isochronous by some method. This prediction model mainly consists of three parts: (1) transform non-isochronous into isochronous, (2) establish nonhomogeneous gray model NGM (1, 1), and (3) inversely transform the isochronous into non-isochronous. These three parts would be given in details as follows:

1. Transform non-isochronous into isochronous
   Define the raw non-isochronous data sequence as

$$X_0^{(0)} = \{x_0^{(0)}(1), x_0^{(0)}(2), \ldots, x_0^{(0)}(k), \ldots, x_0^{(0)}(N)\}, \quad k \geq 1 \quad (6.3)$$

where N is the total number of known data used to establish gray model.
   The time interval of data is

$$T_k = t_{k+1}, -t_k \quad k = 1, 2, \ldots N - 1 \quad (6.4)$$

Thus, its average time interval is

$$\Delta T = \frac{1}{N-1} \sum_{k=1}^{N-1} T_k = \frac{1}{N-1} (t_N - t_1) \quad (6.5)$$

Use cubic spline interpolation method to transform the non-isochronous data sequence (6.3) into the new series (6.6) of time interval ΔT

$$X_1^{(0)} = \{x_1^{(0)}(1), x_1^{(0)}(2), \ldots, x_1^{(0)}(k), \ldots x_1^{(0)}(N)\} \quad (6.6)$$

where $X_1^{(0)}$ is the available isochronous data sequence.

2. Establish nonhomogeneous gray model NGM (1, 1)
   In order to smooth the randomness, the generated non-isochronous data sequence (6.6) is subjected to AGO, and 1-AGO sequence is obtained:

$$X_1^{(1)} = \{x_1^{(1)}(1), x_1^{(1)}(2), \ldots, x_1^{(1)}(k), \ldots x_1^{(1)}(N)\} \quad (6.7)$$

in which $X_1^{(1)}$ is the 1-AGO sequence and $x_1^{(1)}(k) = \sum_{i=1}^{k} x_1^{(1)}(i)$, $k = 1, 2, \ldots N$.
   The generated mean sequence $Z_1^{(1)}$ of $X_1^{(1)}$ is defined as

$$Z_1^{(1)} = \{z_1^{(1)}(1), \ldots, z_1^{(1)}(k), \ldots z_1^{(1)}(N-1)\} \quad (6.8)$$

where $z_1^{(1)}(k)$ is the mean value of adjacent data, i.e., $z_1^{(1)}(k) = \frac{1}{2}(x_1^{(1)}(k)+x_1^{(1)}(k+1))$, $k = 1, 2, \ldots N - 1$.

The least square estimate sequence of the gray differential equation of NGM (1, 1) is defined as follows:

$$x_1^{(0)}(k) + az_1^{(1)}(k) = bk + c \tag{6.9}$$

where a, b, c are model parameters. They could be calculated by the least square method:

$$[a, b, c]^T = (B^T B)^{-1} B^T Y \tag{6.10}$$

in which $B = \begin{bmatrix} -z_1^{(1)}(1) & 3/2 & 1 \\ -z_1^{(1)}(2) & 5/2 & 1 \\ \vdots & \vdots & \vdots \\ -z_1^{(1)}(n-1) & (2n+1)/2 & 1 \end{bmatrix}$, $Y = \begin{bmatrix} x_1^{(0)}(2) \\ x_1^{(0)}(3) \\ \vdots \\ x_1^{(0)}(n) \end{bmatrix}$

It is noteworthy that the gray differential equation of conventional gray model GM (1, 1) is defined as (6.11) (Deng 1989), which is different from (6.9):

$$x_1^{(0)}(k) + az_1^{(1)}(k) = c \tag{6.11}$$

The whitening equation (see Deng 1989 about definition of whitening equation) of (6.9) is therefore as follows:

$$\frac{dx_1^{(1)}(t)}{dt} + ax_1^{(1)}(t) = bt + c \tag{6.12}$$

The solution of (6.12) is

$$x_1^{(1)}(t) = Ce^{-at} + \frac{b}{a}t - \frac{b}{a^2} + \frac{c}{a} \tag{6.13}$$

where C is a parameter, which can be computed by initial conditions.

Consequently, the solution of $x_1^{(1)}(k)$ is

$$\widehat{x}_1^{(1)}(k) = Ce^{-a(k-1)} + \frac{b}{a}k - \frac{b}{a^2} + \frac{c}{a}, \quad k = 1, 2, \ldots, N \tag{6.14}$$

at $k=1$, $\widehat{x}_1^{(1)}(k) = x_1^{(1)}(1) = x_0^{(0)}(1)$; thus,

$$\widehat{x}_1^{(1)}(k) = \left(x_0^{(0)}(1) - \frac{b}{a} + \frac{b}{a^2} - \frac{c}{a}\right)e^{-a(k-1)} + \frac{b}{a}k - \frac{b}{a^2} + \frac{c}{a},$$

$$k = 1, 2, \ldots, N \tag{6.15}$$

Thus, the solution of $x_1^{(0)}(k)$ is

$$\widehat{x}_1^{(0)}(k) = \begin{cases} x_1^{(0)}(1), & k = 1 \\ (1 - e^a)\left(x_0^{(0)}(1) - \frac{b}{a} + \frac{b}{a^2} - \frac{c}{a}\right)e^{-a(k-1)} + \frac{b}{a}, & k = 2, 3, \ldots, N \end{cases} \quad (6.16)$$

The isochronous prediction series designated as $\widehat{x}_1^{(0)}(k')$ could be computed by

$$\widehat{x}_1^{(0)}(k'+1) = (1 - e^a)\left(x_0^{(0)}(1) - \frac{b}{a} + \frac{b}{a^2} - \frac{c}{a}\right)e^{-ak'} + \frac{b}{a},$$
$$k' = N, N+1, N+2, \ldots, N+n \quad (6.17)$$

3. Return the $\widehat{x}_1^{(0)}(k), \widehat{x}_1^{(0)}(k')$ back to $\widehat{x}_0^{(0)}(k)$

Still use cubic spline interpolation method to transform $\widehat{x}_1^{(0)}(k), \widehat{x}_1^{(0)}(k')$ into $\widehat{x}_0^{(0)}(k), \widehat{x}_0^{(0)}(k')$ (the fitting and predicting results of sequence)

$$\widehat{X}_0^{(0)} = \left\{\widehat{x}_0^{(0)}(1), \widehat{x}_0^{(0)}(2), \ldots, \widehat{x}_0^{(0)}(k), \ldots \widehat{x}_0^{(0)}(N)\right\} \quad (6.18)$$

$$\widehat{X}_0'^{(0)} = \left\{\widehat{x}_0^{(0)}(N+1), \widehat{x}_0^{(0)}(N+2), \ldots, \widehat{x}_0^{(0)}(k'), \ldots, \widehat{x}_0^{(0)}(N+n)\right\} \quad (6.19)$$

where $\widehat{x}_0^{(0)}(k), \widehat{x}_0^{(0)}(k')$ are fitting and predicting results, respectively.

So far, the NNGM (1, 1) is built completely. It can be seen from the above derivation that the nonhomogeneous gray model NGM (1, 1) is a special case of nonisochronous nonhomogeneous exponential gray model (NNGM) (1, 1). The NNGM (1, 1) will be equal to the NGM (1, 1) when the modeling data is isochronous.

6.4.3 Model Checking

Any prediction model must be checked up before it is applied. There are three main checking methods by use of residuals, which are average relative error, standard deviation ratio (also known as posterior error ratio), and small error probability (see Liu et al. 2010). Reference level of model accuracy is shown in Table 6.3:

1. Average relative error

$$\alpha = \frac{1}{n}\sum_{k=1}^{n}\left|\frac{\varepsilon(k)}{x_0^{(0)}(k)}\right| \quad (6.20)$$

where $\varepsilon(k)$ are residuals of model $\varepsilon(k) = x_0^{(0)}(k) - \widehat{x}_0^{(0)}(k)$

Table 6.3 Reference level of model accuracy (Liu et al. 2010)

Accuracy class	Average relative error a	Standard deviation ratio C	Small error probability p
I (Excellent)	$0.00 \leq \alpha < 0.01$	$0.00 \leq C < 0.35$	$1 \geq p \geq 0.95$
II (Good)	$0.01 \leq \alpha < 0.05$	$0.35 \leq C < 0.50$	$0.95 > p \geq 0.80$
III (Satisfactory)	$0.05 \leq \alpha < 0.10$	$0.50 \leq C < 0.65$	$0.80 > p \geq 0.70$
IV (Failed)	$0.10 \leq \alpha < 0.20$	$0.65 \leq C < 0.80$	$0.70 > p \geq 0.60$

2. Standard deviation ratio

$$C = \frac{S_2}{S_1} \quad (6.21)$$

in which S_1, S_2 are the standard deviations of $X_0^{(0)}$ (the primitive data sequence) and $\widehat{X}_0^{(0)}$ (the corresponding predicted result sequence), respectively. They are defined as follows:

$$S_1 = \sqrt{\frac{1}{n}\sum_{k=1}^{n}\left[x_0^{(0)}(k) - \overline{x}\right]}, \quad \overline{x} = \frac{1}{n}\sum_{k=1}^{n} x_0^{(0)}(k),$$

$$S_2 = \sqrt{\frac{1}{n}\sum_{k=1}^{n}\left[\varepsilon(k) - \overline{\varepsilon}\right]}, \quad \overline{\varepsilon} = \frac{1}{n}\sum_{k=1}^{n} \varepsilon(k)$$

3. Small error probability

$$p = P\left(|\varepsilon(k) - \overline{\varepsilon}| < 0.6745 S_1\right) \quad (6.22)$$

6.4.4 Applications of Prediction Model

To verify the applicability and prediction accuracy of NNGM (1, 1), some available laboratory test results and in-site monitoring results of cumulative plastic strain in literatures are compared with their corresponding predicted results calculated by this method.

6.4.4.1 Application in Laboratory Experiments

Dynamic laboratory test results of cumulative plastic strain of different soils are used to evaluate the NNGM (1, 1). These soils include saturated and unsaturated soft clay, undisturbed and reconstituted clay, and fine-grained foundation soils and sand (showed in Fig. 6.9). The results are showed in Fig. 6.10 and Table 6.4. It could be seen from them that both standard deviation ratio C and small error probability

6.4 Applications of Gray Prediction Theory in Predicting Long-Term... 211

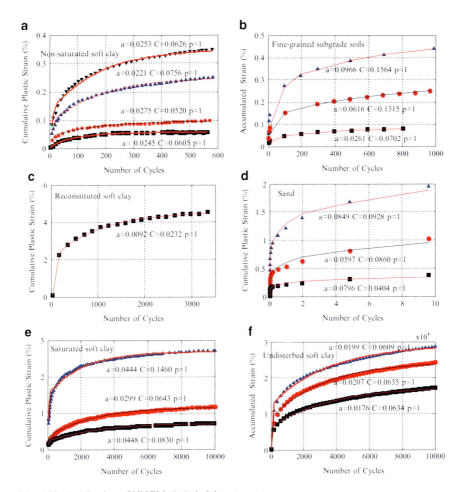

Fig. 6.10 Application of NNGM (1, 1) in laboratory tests

p of this model in all these soils are in class I (excellent) and that of the average relative error α is above III (satisfactory). The biggest relative error in all these prediction results is no more than 10 % (9.66 % in Li and Selig's test), and the smallest is only about 1 % (0.92 % in Buddhima's test). Thus, it could be believed that the improved gray model NNGM (1, 1) proposed here is a promising method to predict the cumulative plastic strain in laboratory tests, and it is valid for a wide range of soils.

6.4.4.2 Application In-Site Monitoring

The maximum settlement of Line 4, Shanghai Metro has come up to 160 mm since the operation in September 2005. Figure 6.11 shows the monitoring results of tunnel

Table 6.4 Accuracy class of NNGM (1, 1)

	α (%)	C	p	Accuracy	Sources
a	2.45	0.0605	1	II	Liu et al. (2008)
	2.75	0.0756	1	II	
	2.21	0.0756	1	II	
	2.53	0.0626	1	II	
b	2.61	0.0702	1	II	Li and Selig (1996)
	6.16	0.1315	1	III	
	9.66	0.1564	1	III	
c	0.92	0.0232	1	I	Buddhima et al. (2009)
d	7.96	0.0404	1	III	Wichtmann et al. (2010)
	5.97	0.0860	1	III	
	8.49	0.0928	1	III	
e	4.48	0.0830	1	II	Huang et al. (2006)
	2.99	0.0643	1	II	
	4.44	0.1460	1	II	
f	1.76	0.0634	1	II	Liu (2008)
	2.07	0.0635	1	II	
	1.99	0.0609	1	II	

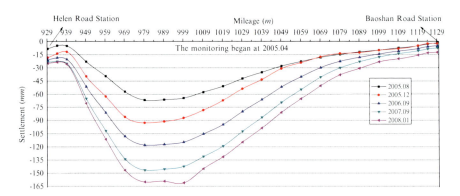

Fig. 6.11 Monitoring data of tunnel settlement of Metro Line 4 from Helen Road Station to Baoshan Road Station in Shanghai (Wang 2011; Ren et al. 2012)

settlement of Line 4, Shanghai Metro from Helen Road Station to Baoshan Road Station conducted by Shanghai Shen Yuan Geotechnical Engineering Co., Ltd. The settlement is mainly due to the cumulative plastic deformation of foundation soils caused by cyclic loading of subway. And deformations of the third layer (mucky silty clay with sandy silt), fourth layer (silt clay), and fifth layer (gray clay with silty clay) of strata have much greater contribution to it. And reference properties are shown in Table 6.5.

It can be seen from Fig. 6.11 that the monitoring consisted of five curves form April 2005 to January 2008. We used the former four to build NNGM (1, 1) and

6.4 Applications of Gray Prediction Theory in Predicting Long-Term... 213

Table 6.5 Reference properties of subgrade soils of Metro Line No. 4 in Helen Road Station

Soil properties	③	④	⑤
Depth (m)	2.63	8.25	18.56
Moisture content (%)	36.0–48.8	41.2–58.0	30.5–40.7
Void ratio	1.02–1.37	1.17–1.67	0.86–1.09
Density (g/cm^3)	1.93	1.76	1.81
Specific gravity	2.73	2.75	2.73
Degree of saturation (%)	92–100	94–99	88–99
Angle of internal friction (°)	18.6–27.7	11.5–16.4	10.0–17.5
Cohesion (kPa)	5.6–7.8	11.3–15.2	9.0–14.1
Compression modulus (MPa)	2.18–2.42	2.03–3.01	2.47–7.46

predicted the fifth curve to verify the established model. The fitted values (the former four) and predicted values (the fifth) are shown in Table 6.6 as well as the model accuracy and the relative error of predicted results. For fitting accuracy of the model, standard deviation ratio C and small error probability are excellent for all of the monitoring points; average relative error α is satisfactory for most of the monitoring points except for the fifth of them, whose $\alpha \geq 10\%$. For predicting accuracy of the model, the relative error of most of the monitoring points is less than 15 % except for seven of them. The average relative error for all of the monitoring points is 11.04 %, which is good enough for prediction in practical engineering applications.

6.4.5 Discussions and Error Analyses

In this section, we compare the proposed model NNGM (1, 1) with traditional gray model GM (1, 1) and nonhomogeneous exponential model (NEM). In addition, error sources of NNGM (1, 1) are discussed:

6.4.5.1 Comparison of GM (1, 1) and NNGM (1, 1)

Although GM (1, 1) is the most widely used prediction model, it can't be used directly to predict non-isochronous series data. NNGM (1, 1) is aimed to expand the scope of applicability of the gray model and improve its precision. To test performance of the new model, one laboratory experiment data (Fig. 6.9, provided by Buddhima et al. 2009) in Fig. 6.9 and one in-site monitoring point data (at the mileage of 979 m) in Fig. 6.11 are selected for predicting the two models, respectively. And the comparison results are shown in Fig. 6.12. Figure 6.12 show the fitted results for laboratory experiment data with NNGM (1, 1) and GM (1, 1), respectively. The average relative error α and the standard deviation ratio C of NNGM (1, 1) are both smaller than those of GM (1, 1). Figure 6.12 show the prediction results of NNGM (1, 1) and GM (1, 1) for the fifth monitoring at

Table 6.6 Predicted results of tunnel settlement based on in-site monitoring data

Mileage		Monitoring recordings Fitted 2005.08	2005.12	2006.09	2007.09	Predicted 2008.01	Accuracy of model Fitted α (%)	C	p	Predicted α' (%)
929	Monitored	−8.49	−18.42	−21.1	−23.53	−24.85	11.93	0.17	1	9.92
	NNGM (1.1)	−12.70	−18.28	−20.36	−22.16	−22.38				
934	Monitored	−6.88	−13.81	−18.44	−22.71	−23.59	7.26	0.10	1	2.48
	NNGM (1.1)	−9.03	−13.82	−18.03	−22.09	−23.00				
939	Monitored	−5.09	−11.83	−20.42	−25.08	−25.97	6.39	0.06	1	1.89
	NNGM (1.1)	−6.54	−12.06	−20.36	−24.77	−25.48				
949	Monitored	−22.93	−39.67	−51.44	−64.95	−70.67	4.35	0.06	1	4.09
	NNGM (1.1)	−26.79	−39.89	−50.10	−63.83	−67.78				
959	Monitored	−39.47	−62.55	−81.24	−101.91	−111.56	2.23	0.03	1	4.64
	NNGM (1.1)	−42.23	−63.14	−79.70	−100.64	−106.38				
969	Monitored	−57.05	−85.95	−108	−133.91	−146.78	1.57	0.02	1	4.57
	NNGM (1.1)	−59.08	−86.91	−105.82	−132.42	−140.07				
979	Monitored	−66.66	−92.7	−117.99	−146.63	−160.12	1.66	0.02	1	3.83
	NNGM (1.1)	−63.48	−94.12	−116.56	−145.89	−153.99				
989	Monitored	−66.21	−91.18	−117.2	−145.4	−159.28	1.88	0.03	1	4.24
	NNGM (1.1)	−62.14	−92.68	−116.04	−144.85	−152.52				
999	Monitored	−64.35	−87.13	−114.6	−142.25	−161.19	2.30	0.04	1	7.44
	NNGM (1.1)	−58.76	−88.76	−113.91	−142.11	−149.20				
1,009	Monitored	−57.66	−78.32	−105.33	−131.11	−145	2.40	0.05	1	5.17
	NNGM (1.1)	−52.25	−79.88	−104.90	−131.14	−137.50				
1,019	Monitored	−50.99	−67.26	−94.73	−119.2	−131.42	3.37	0.07	1	4.34
	NNGM (1.1)	−43.97	−68.90	−94.83	−119.86	−125.72				
1,029	Monitored	−42.26	−53.79	−79.87	−102.39	−114.69	4.68	0.08	1	5.00
	NNGM (1.1)	−34.35	−55.37	−80.36	−103.55	−108.96				

6.4 Applications of Gray Prediction Theory in Predicting Long-Term...

1,039	Monitored	−35.07	−43.35	−66.3	−86.56	−98.74				
	NNGM(1,1)	−27.33	−44.73	−66.92	−87.81	−92.81	5.53	0.10	1	6.00
1,049	Monitored	−28.01	−30.88	−51.78	−69.59	−80.93				
	NNGM(1,1)	−18.66	−32.28	−52.81	−71.45	−75.92	8.51	0.14	1	6.19
1,059	Monitored	−22.86	−24.19	−40.19	−54.68	−65.46				
	NNGM(1,1)	−14.75	−25.29	−41.09	−56.21	−59.98	9.01	0.16	1	8.37
1,069	Monitored	−18.1	−17.52	−29.82	−40.64	−50.54				
	NNGM(1,1)	−10.63	−18.49	−30.66	−42.11	−44.94	10.65	0.20	1	11.08
1,079	Monitored	−14.74	−13.73	−22.62	−30.07	−38.13				
	NNGM(1,1)	−8.53	−14.54	−23.27	−31.30	−33.24	11.00	0.22	1	12.83
1,089	Monitored	−11.88	−12.66	−18.03	−23.07	−30.64				
	NNGM(1,1)	−8.37	−13.15	−18.30	−23.61	−24.93	7.44	0.17	1	18.63
1,099	Monitored	−9.83	−10.07	−14.26	−17.78	−23.34				
	NNGM(1,1)	−6.70	−10.52	−14.49	−18.31	−19.20	8.16	0.19	1	17.74
1,109	Monitored	−7.37	−8.37	−11.05	−13.89	−19.78				
	NNGM(1,1)	−5.70	−8.61	−11.12	−14.08	−14.85	5.51	0.13	1	24.90
1,119	Monitored	−4.04	−4.86	−8.13	−10.77	−15.69				
	NNGM(1,1)	−2.92	−5.06	−8.25	−11.00	−11.63	7.10	0.11	1	25.85
1,124	Monitored	−1.99	−2.52	−5.99	−9.72	−13.14				
	NNGM(1,1)	−1.13	−2.61	−6.16	−9.91	−10.98	10.38	0.09	1	16.46
1,129	Monitored	−1.82	−2.38	−5.1	−8.52	−12.39				
	NNGM(1,1)	−1.19	−2.43	−5.20	−8.63	−9.71	8.00	0.08	1	21.66
1,134	Monitored	−1.67	−1.83	−4.08	−6.30	−11.35				
	NNGM(1,1)	−0.89	−1.92	−4.21	−6.47	−7.08	11.59	0.13	1	37.64

Note: α' is relative error. $\alpha' = |(x_{\text{tested}} - x_{\text{monitered}})/x_{\text{tested}}|$

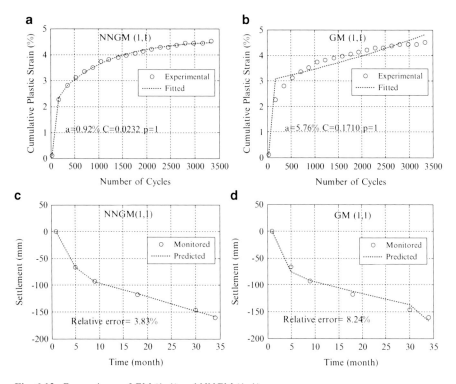

Fig. 6.12 Comparisons of GM (1, 1) and NNGM (1, 1)

distance of 979 m. The relative errors of these two models are 3.83 % and 8.24 %, respectively. It indicates that the NNGM (1, 1) has a much better performance than GM (1, 1), not only in fitting cumulative plastic strain of laboratory experiments but also in predicting long-term settlement of the subway tunnel.

6.4.5.2 Comparison of NEM Fitting and NNGM (1, 1)

Although both the NEM and NNGM (1, 1) are capable to predict the nonhomogeneous exponential series, there is essential difference between them. While raw data without any processing is used to build NEM, 1-AGO data generated by adding raw data in series successively is used for NNGM (1, 1). The accumulated generating operation (AGO) enhances the regularity of the raw data, which is useful for improving accuracy of models. Equation (6.23) is going to be used to establish an NEM:

$$y = ae^{bx} + c \tag{6.23}$$

6.4 Applications of Gray Prediction Theory in Predicting Long-Term... 217

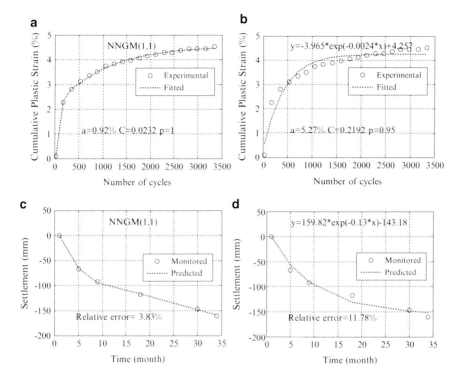

Fig. 6.13 Comparison of NNGM (1, 1) and NEM

where a, b, c are fitting parameters, which can be calculated by the least square method.

The same data given in Fig. 6.12 are used to build the two prediction models. The fitted and predicted results are shown in Fig. 6.13. It also shows a better performance of NNGM (1, 1) than NEM's.

6.4.5.3 Error Source of NNGM (1, 1)

The error source of the proposed model mainly consists of three parts. One is inherent deviation of model because its gray differential equation (6.9) does not exactly match with its whitenization differential equation (6.12). This error could be eliminated by establishing unbiased model using some methods. However, no matter what method is used in the current study, the cost of that is greatly increasing the complexity of forecast. It should be noted that the computational burden is the most important parameter after the performance in real-time applications. Therefore, it doesn't seem to be worth exchanging efficiency for a small improvement of prediction accuracy.

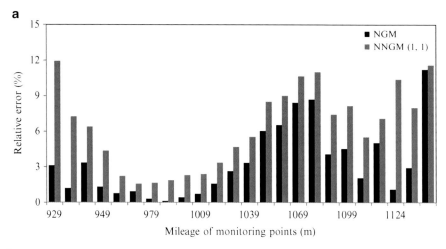

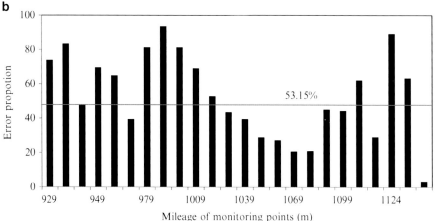

Fig. 6.14 Relative errors of NNGM (1, 1) and NGM (1, 1)

The second error source is the interconversion process of the data series between non-isochronous and isochronous. This error does not exist if the modeling data are isochronous. Therefore, the proportion of this part error to total error can be calculated by using the same modeling data to build NNGM (1, 1) and NGM (1, 1) (see Sect. 6.4.2 in detail):

$$\Delta = \frac{a_{\text{NNGM (1,1)}} - a_{\text{NGM (1,1)}}}{a_{\text{NNGM}}} 100 \qquad (6.24)$$

The monitoring data in Fig. 6.11 are used to compare relative error of these two models, and the comparison results are shown in Fig. 6.14. Figure 6.14 shows the relative error of NNGM (1, 1) is always larger than NGM (1, 1), which demonstrates the interconversion process of NNGM (1, 1) between non-isochronous and isochronous generate error. Figure 6.14 shows the proportion of error generated from the interconversion to the total error. The average proportion of all monitoring points reaches up to 53 %. It indicates that the interconversion process of non-isochronous to isochronous is the main error source of NNGM (1, 1). Therefore, how to reduce this part of error is the key to improve the accuracy of NNGM (1, 1), and it is one aspect of our future research work.

In addition, the third part of error comes from the model itself. Even if the unbiased model has been established to forecast isochronous data series, model error won't be completely avoided because the information of predicted objects is not fully known. Thus, it is futile to improve accuracy of model by trying to reduce this part of error.

Gray prediction model is a reasonable and potential approach at least in theory to simulate and predict cumulative plastic deformation of soils under cyclic loads. The proposed gray model NNGM (1, 1) evaluated by numerous experimental data of existing literatures was found to be valid for a wide range of soils. It has an excellent performance in predicting long-term settlement of the subway. The error of this model is mainly generated by the interconversion process between non-isochronous data sequence and isochronous ones. Therefore, a good transformation method is the key point to improve the prediction accuracy of the model.

6.5 Effective Stress Analysis of Long-Term Settlement in Tunnel

For analyzing the vibration subsidence of soil under the long-term train vibration load, effective stress method is generally applied. Before analyzing a problem, we should determine whether dynamic analysis is necessary and whether the dissipation and diffusion of pore water pressure appears under dynamic load. The field monitoring data shows that the frequency of soil vibration induced by the subway is quite low and the duration of the vibration is short. Therefore, the residual excess pore water pressure is not great, and it has enough time to dissipate. For homogeneous soil, the speed of cyclic load can be evaluated by the following two dimensionless parameters:

$$\Pi_1 = \pi^2 \left(\frac{T_0}{T}\right)^2 \qquad (6.25)$$

where T_0 denotes the natural period of vibrating system and T is the period of dynamic load. Equation (6.25) is applied for evaluating the speed of cyclic loads. Generally, it is not necessary to process dynamic analysis if Π_1 is less than 10^{-3}. If we consider the pass of one subway train as one applied loading, the subway operation interval could be considered as the period. This interval of subway train is generally 3–5 min; in other words, the loading period is more than 180 s. For soft

Fig. 6.15 Ground settlement curve from March 2000 to December 2002

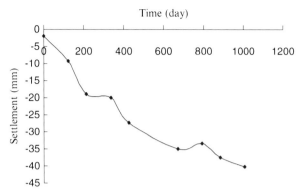

soil in Shanghai, the natural period is $T_0 < 0.5$ s and $\Pi_1 = 7.6 \times 10^{-5} < 10^{-3}$. In this case, each subway train passing could be considered as a process of quasi-static loading.

Experimental studies have shown that vibration subsidence tests are very complex; the primary contradiction is the irreconcilability between the requirements of a large number of experiments with the inhomogeneity of undisturbed soil sample. As everyone knows, the real soil layer is anisotropic and heterogeneous; properties of soils are quite different from each other due to various formation environments and geological times. Even for the same soil layer, specimens obtained from it also vary due to the interference of drilling, transportation, cutting, and other external factors. This phenomenon is more distinct in soft soil. Therefore, it is very difficult to guaranty the uniformity of all the tested specimens. Generally, every group vibration subsidence test needs at least 30 effective specimens; the test results are always very discrete. Because of this, all of the data obtained from field tests and monitoring are normalized in this chapter.

Figure 6.15 shows the curves of ground settlement monitored from March 2000 to December 2002. Its horizontal axis is monitoring time.

Subway operation time is from 5:30 to 23:00 every day. The operating interval is 5 min, and the train number in 1 day is about 210 times. Then we applied the residual deformation induced by single passing train which was obtained from simulated computation to fit the curves of on-site monitoring subsidence; the fitting equation was drawn as follows:

$$S = -\Delta S^{C_1} \cdot (\ln(x + C_2))^{C_3} \qquad (6.26)$$

where S (mm) denotes ground subsidence; ΔS denotes residual deformation induced by single passing train; $\Delta S = 0.121$ mm was computed by numerical simulation; x is the number of passing trains; C_1, C_2, C_3 are fitted parameters; and C_1, C_2, C_3 could be given by fitting the monitoring data showed in Fig. 6.16, $C_1 = 6.043$, $C_2 = 866.752$, $C_3 = 5.567$. The coefficient of correlation is 0.983,

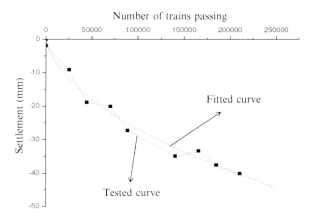

Fig. 6.16 Comparison chart of fitting curve and measured curve

which indicates there is a good correlation between settlement and the number of passing trains.

Equation (6.26) can be used to compute and predict the long-term settlement of subway. For example, the calculated settlement is 78 mm in 2010. It is worth noting that the subway settlement is determined by many factors, while two of them should be paid more attention. One is the characteristics of subway itself including the vehicle axle load and wheel flat and track irregularities. The other one is the effect of engineering constructions adjacent to the subway tunnel. Because there are few constructions around our test site, the first factor should be the dominant one.

6.6 Chapter Summary

This chapter made a comprehensive analysis of the influence factors on subway settlement and proposed a prediction model on the basis of gray system theory:

1. In soft soil area, ground settlement caused by tunnel excavation is controlled by many factors such as grouting, volume of excavated earth, and the soil pressure and advancing speed. For controlling the influence of tunnel excavation on the surrounding environment, it is necessary to take all the key factors into consideration to make appropriate construction projects.
2. Grouting volume and slurry consistency are crucial factors to control ground settlement. Considering the heave effect on the tunnel axis, the volume of grouting should be between 2.5 and 2.8 m^3 when the slurry consistency is between 9 and 11.
3. The non-isochronous nonhomogeneous exponential gray prediction model NNGM (1, 1) was improved based on a traditional model GM (1, 1). We reconstructed the whitenization differential equation of GM (1, 1) to expand its application and predict nonhomogeneous exponential data series. To meet the requirements of gray model of the raw data, non-isochronous data sequence were

transformed into the isochronous by use of cubic spline interpolation and then transformed back by the same method. In addition, we analyzed the long-term settlement of the subway tunnel in the aspect of effective stress in the last part of this chapter.
4. Gray prediction model is a reasonable and potential approach to simulate and predict cumulative plastic deformation of soils under cyclic loads and long-term settlement of subway tunnel.
5. Compared with gray prediction theory, the effective stress method is more like an empirical predicted model for predicting the settlement of subway. It is more suitable to make a simple qualitative analysis and evaluation for deformation and settlement. The gray model could be used to predict the long-term settlement.

References

Brown SF (1974) Repeated load testing of a granular material. J Geotech Eng Div, ASCE 100(7):825–841

Chai JC, Miura N (2002) Traffic-load-induced permanent deformation of road on soft subsoil. J Geotech Geoenviron Eng 128(11):907–916

Chen JW, Zhan LG (2000) Deformation measuring of the metro tunnel and deformation data analysis of shanghai metro line No. 1. Shanghai Geol 2:51–56 (in Chinese)

Dafalias YF (1986) Bounding surface plasticity (I): mathematical formulation and hypoplasticity. J Eng Mech ASCE 112(9):966–987

Dafalias YF, Herrmann LR (1986) Bounding surface plasticity (II): application to isotropic cohesive soils. J Eng Mech ASCE 112(12):1263–1291

Dai BH (2008) Prediction of the effect of deep excavation on adjacent metro tunnel. Study City Track Traffic 8(1):62–65

Deng JL (1982) The control problem of grey systems. Syst Control Lett 1(5):288–294

Deng JL (1989) Introduction to grey system theory. J Grey Syst 1:1–24

Huang MS, Li JJ, Li XZ (2006) Cumulative deformation behaviour of soft clay in cyclic undrained tests. Chin J Geotech Eng 28(7):891–895 (in Chinese)

Hyde AFL, Brown SF (1976) The plastic deformation of a silty clay under creep and repeated loading. Geotechnique 26(1):173–184

Hyodo M, Yasuhara K, Htrao K (1992) Prediction of clay behaviour in undrained and partially drained cyclic tests. Soils Found 32(4):117–127

Indraratna B, Attya A, Rujikiatkamjorn C (2009) Experimental investigation on effectiveness of a vertical drain under cyclic loads. J Geotech Geoenviron Eng 135(6):835–839

Kayacan E, Ulutas B, Kaynak O (2010) Grey system theory-based models in time series prediction. Expert Syst Appl 37:1784–1789

Li DQ, Selig ET (1996) Cumulative plastic deformation for tine grained subgrade soils. J Geotech Eng 122(12):1006–1013

Liao SM, Hou XY (1998) The latest progress in design and construction of soft soil shield tunnel. Undergr Space 18(5):406–412 (in Chinese)

Liu JH, Hou XY (1991). Shield driven tunnel [M]. Beijing, China Railway Publishing House (in Chinese)

Liu S (2008) Study on the rheological properties of saturated soft clay around the tunnel under subway loading. Master thesis, Tongji University, Shanghai

Liu FC, Shang SP, Wang HD (2008) Study of strain accumulation strengthened model for clay under cyclic loadings. Rock Soil Mech 29(9):2457–2461 (in Chinese)

Liu SF, Dang YG, Fang ZG, Xie NM (2010) Grey system theory and its application, 5th edn. Science Press, Beijing

Miura N, Fujikawa K, Sakai A, Hara K (1995) Field measurement of settlement in Saga airport highway subjected to traffic load. Tsuchi-to-Kiso 43–6(449):49–51

Monismith CL, Ogawa N, Freeme CR (1975) Permanent deformation characteristics of subgrade soils due to repeated loading. Transportation Research Board, Washington, DC, pp 1–17

Prevost JH (1978) Plasticity theory for soils stress-strain behavior. J Eng Mech Div ASCE 104(5):1177–1196

Ren XW, Tang YQ, Li Q, Yang Q (2012) A prediction method using grey model for cumulative plastic deformation under cyclic loads. Nat Hazards 64:441–457

Sakai A, Samang L, Miura N (2003) Partially-drained cyclic behavior and its application to the settlement of a low embankment road on silty-clay. Soils Found 43(1):33–46

Sun JL (2006) Monitoring and analysis of settlement of Metro Tunnel. Master's degree thesis of Hehai University, Nanjing, China

Tang YQ, Ye WM, Zhang QH (1993) Study on some problems of the construction of shield tunnel in Shanghai Metro (2) Mechanism analysis of shield tunnel axis deviation. Undergr Space 13(3):171–177 (in Chinese)

Tang YQ, Zhou J, Liu S, Yang P, Wang JX (2011) Test on cyclic creep behavior of mucky clay in Shanghai under step cyclic loading. Environ Earth Sci 63:321–327

Tong F (2000) Analysis of metro tunnel and seismic dynamic of joint structure. Ph.D. thesis, Tongji University, Shanghai, China

Wang YD (2011) The study on the dynamic characteristic and long-term settlement of reinforced soft clay around the tunnel under subway vibration loading [D]. Tongji University

Wang HM, Tang YQ, Yan X et al (2006) Prediction of the land subsidence in soft soil area of non-isochronous GM (1, 1) model [J]. J Eng Geol 14(3):398–400

Wang SP, Yan GX, Xie JF (2001) Seismic liquefaction analysis of Nanjing subway line Xu Fu Xiang NanJing Railway Station shield interval. Modern Tunn Technol 38(2):19–23 (in Chinese)

Wichtmann T, Rondón HA, Niemunis A, Triantafyllidis T, Lizcano A (2010) Prediction of permanent deformations in pavements using a high-cycle accumulation model. J Geotech Geoenviron Eng 136(5):728–740

Yu SS, Shi ZJ, Xie JF et al (1986) Calculation and analysis of Shanghai subway tunnel seismic subsidence. Earthq Eng Eng Vib 6(3):51–59

Zheng YL, Pan J, Han WX (2005) Settlement analysis of metro tunnels in soft soil. Chin J Underg Space Eng 1(1):16–19 (in Chinese)

Zhou QN, Wang X (2002) The application of gray theory in the prediction of settlement. Soil Eng Found 16(4):31–33 (in Chinese)

Zienkiewicz OC, Leung KH, Pastor M (1985) Simple model for transient soil loading in earthquake analysis (I): basic model and its application. Int J Numer Anal Methods Geomech 9:453–476

Chapter 7
Conclusions and Prospects

7.1 Conclusions

In this monograph, we took Shanghai muddy clay as the study object and analyzed the characteristics of soil dynamic response under subway loadings based on field monitoring of earth pressure and pore water pressure within soils. Benefiting from the field test, some basic parameters were obtained for the laboratory tests. A series of laboratory tests were conducted with a dynamic triaxial apparatus with unidirectional excitation combined with a multifunctional dynamic triaxial test system. The research focused on four main aspects: (1) the increasing law of pore water pressure of soft clay under subway load, (2) the relationship between dynamic stress and strain, (3) the law of strength degradation under dynamic loading, and (4) creep behavior under dynamic loading. In addition to the experimental study above, 3D finite element model was established to simulate dynamic response of soil surrounding the subway tunnel when the subway trains goes through. Besides the macroscopical study of the research object, the microstructure of soil was studied as well based on mercury intrusion porosimetry (MIP) and scanning electron microscopy (SEM). The former one provides access to quantitatively analyze the microstructure of soft clay before and after vibration (such as variations of pore size distribution, pore quantity, and other microcosmic parameters), while the images of SEM provided visible and qualitative analysis of soil structure. The mechanism of soil deformation can be revealed based on these two methods. Some beneficial conclusions were drawn as follows:

1. Field monitoring data shows that (1) the frequency of soft clay responding to train operation can be divided into two parts: high frequency (2.4–2.6 Hz) and low frequency (0.4–0.6 Hz); the amplitude of the dynamic responding stress varies from depth as the responding stress amplitudes of 8.5, 11.5, and 13.5 are 0.23 kPa, 0.70 kPa, and 1.15 kPa, respectively. The formula which demonstrates the degradation of dynamic response of soil in the perpendicular direction to the tunnel axis was acquired; and a linear relationship exists between the amplitude

of dynamic responding stress and the depth. (2) The response of pore water pressure has an obvious hysteresis effect to subway loading; and the smaller distance from subway tunnel leads to shorter hysteresis time. The period of pore pressure fluctuation is basically equal to the subway vibration period. (3) The response of pore water pressure with smaller distance from subway tunnel is much more sensitive. And soil underlying the subway tunnel is more sensitive than the overlying. The vibration transmission stops at a certain depth when the vibration energy is consumed by the pore water motion fully.

2. The growth pattern of pore water pressure under subway loading in saturated soft clay surrounding the tunnel can be divided into three stages: rapid growth stage, slow growth stage, and stabilization stage. In the rapid growth stage, the increasing rate of pore water pressure is not constant; it grows rapidly in a very short time and then slowly attenuates. The excess pore water pressure decreases rapidly and finally keeps stable at a certain value which is a little higher than the hydrostatic pressure.

3. The cyclic stress ratio, confining pressure, vibration frequency, and vibration cycles play significant roles in deformation characteristics. There is a critical dynamic stress ratio which is determined by the nature properties of soil, dynamic loading, and consolidation confining pressure. More specifically, when the applied load is smaller than the critical dynamic stress, though soil deformation increases with the growth of vibration cycles, the increasing rate and vibration amplitude have a downward trend and tend to be stable. When the applied load is larger than the critical dynamic stress, the deformation grows dramatically until the soil is damaged. In terms of subway tunnel, the deformation of the soil near tunnel sidewalls is relatively small, while the deformation of soil under subway tunnel is larger. In conclusion, during a long period of subway operation, the soil accumulated deformation cannot be neglected in spite of a relatively small deformation during a short time.

4. The 3D finite element model is established according to the data of investigation and field monitoring. During the process of finite element discretization, artificial boundary is selected to simulate the surrounding soil in order to acquire accurate results without too large calculation area. The calculation results were coincident to the corresponding monitoring data; therefore, the model was proved reliable. The results show that the dynamic response mainly concentrates on the soil surrounding subway tunnel and the vertical transmission distance is larger than the horizontal one. At last, the empirical formula for settlement prediction is proposed by fitting the residual deformation under one single subway train with the field-monitored settlement curve.

5. The saturated mucky clay of the fourth layer in Shanghai mainly possesses flocculation structure and honeycomb-flocculation structure. Clay mineral particles mainly consist of illite, less chlorite, and least montmorillonite, etc. The particle is mainly laminar, while the aggregate is flocculent with flower or feather shape. The connections of particles are mainly edge-to-face and edge-to-edge which forms the aerial open structure with high void ratio. The saturated mucky clay of Shanghai consists of a great deal of macropore and has bottleneck effect

7.1 Conclusions

in the process of mercury intrusion. The variation of pore structure parameters with different depths can be indicated by CSR. With the increase of depth, the CSR grows while the specific surface area decreases at first and then rebounds marginally. The mean distribution radius, retention factor, and porosity have the opposite trend.

6. Based on the fractal theory, the study on clay pore distribution shows that Menger sponge random model is not appropriate to describe the fractal characteristic of muddy clay in the whole particle size range, while the model based on thermodynamic relations is suitable. The relationship between fractal dimension calculated by thermodynamic relations model and CSR provides reasonable explanation of variation characteristics. In addition, the microscopic test results also indicate that packing model is not suitable for explaining the mechanism of deformation of saturated soft clay under low stress conditions.

7. In essence, the macroscopic deformation characteristic of saturated soft clay under subway loading is a reconstruction process of microstructure. The correlation analysis of pore microstructure parameters and macroscopic deformation, macroscopic force, shows that a critical value of subway loading plays an important role in the change of microstructure of mucky clay near the subway tunnel. Predictably, during a long time of subway operation, soil structure elements of soils near the subway tunnel and at the bottom of rail have a trend to be compacted. The pressure applied on soil at the bottom of the subway tunnel is quite large which makes the compacted trend less important, while the rearrangement of soil elements is the main reason of deformation. Therefore, the generation of deformation will continue for a long time.

8. The process of soil deformation surrounding the subway tunnel and ground settlement caused by the construction of subway can be divided into two stages. One part is generated during tunnel construction and the other part is caused by subway operation. The ground settlement caused by excavating subway in soft soil layer mainly depends on the quantity of grouting, excavated volume, soil pressure, and advance rate of tunnel excavation. Because the grouting quantity and slurry consistency is the key factor to control ground subsidence, the most proper grouting quantity is 2.5–2.8 m^3 when the slurry consistency is within 9–11. In addition, the stability ratio should be controlled in the range of 5–6. Besides the effect of stratum damage and building gap filling, influence of segment ring deformation and other factors should be taken into account in order to control ground settlement.

9. Gray prediction theory was introduced, and it was proved to be a reasonable and potential approach to simulate and predict cumulative plastic deformation of soils under cyclic loads and long-term settlement of subway tunnel. By applying of cubic spline interpolation to transform raw non-isochronous data to the required isochronous data for gray prediction theory and reconstructing the whitenization differential equation of GM (1,1) to expand it to be capable of predicting nonhomogeneous exponential data series, a non-isochronous nonhomogeneous exponential gray prediction model NNGM (1,1) was proposed to predict cumulative plastic deformation of soils under cyclic loads. It also has a good performance in predicting long-term settlement of subway.

7.2 Prospects for Further Study

In spite of a series of meaningful and practical conclusions, lots of further study is prospected due to the limitation of our research time and research level.

1. Based on the present achievements, further study could focus on the dynamic response and microstructure variation of saturated soft clay when subjected to different loading frequencies. Finally the mechanism model of microstructure deformation can be established for soft clay.
2. In this monograph, the dynamic response of soil under subway loading is calculated using numerical simulation. In the next stage, with the application of a computerized dynamic program, soil dynamic model could be established considering mechanism to verify the results of field test and laboratory test. Then, in terms of improving the stereo disaster prevention system of the urban area, database of relevant parameters of Shanghai Subway could be established to monitor the damage caused by subway long-term operation.
3. Soil pore which is assumed to be smooth cylindrical shape is actually curvilinear with complex unsmooth surface. Some scholars used four times Koch curve to evaluate the pore characteristics and determined the similarity fractal dimension. Therefore, further study on microstructure characteristics of soft clay could be conducted with application of fractal theory and morphology theory.

Appendix: Major Published Works of the Book Author

English Papers

1. **Yiqun Tang**, Weimin Ye, Yu Huang (2003) Marsh gas in shallow soils and safety measures for tunnel construction. Engineering Geology 67(1):373–378.
2. Nianqing Zhou, **Yiqun Tang**, Heping Tang (2005) Groundwater waves in a coastal fractured aquifer of the third phase Qinshan nuclear power engineering field. Journal of Shanghai Jiaotong University (Science) 10(4):441–445.
3. **Yiqun Tang**, Shukai Zhao, Yu Huang, Jianxiu Wang and Nianqing Zhou (2006) Experimental studies on consolidation law of the stratum containing high-pressure gas. ASCE, Geotechnical Special Publication 148:315–322.
4. Nianqing Zhou, **Yiqun Tang**, Yonghuang Deng, Zaili Zhao (2006) Structural characteristics and mechanical properties of rock mass in the field of Tianwan nuclear power plant, China. Journal of Shanghai Jiaotong University (Science) 11(4):512–517.
5. **Yiqun Tang**, Ping Yang, Shukai Zhao, Xi Zhang, Jianxiu Wang (2008) Characteristics of deformation of saturated soft clay under the load of Shanghai subway line No.2. Environmental Geology 54(6):1197–1203.
6. **Yiqun Tang**, Zhengdong Cui, Jianxiu Wang, Xuexin Yan (2008) Application of grey theory-based model to prediction of land subsidence due to engineering environment in Shanghai. Environmental Geology 55(3):583–593.
7. **Yiqun Tang**, Zhengdong Cui, Jianxiu Wang, Chen Lu, Xuexin Yan (2008) Model test study of land subsidence caused by the high-rise building group. Bulletin of Engineering Geology and the Environment 67(2):173–179.
8. Zhengdong Cui, **Yiqun Tang**, Changqing Guo, Li Yuan, Chunling Yan (2008) Flow-induced vibration and stability of an element model for parallel-plate fuel assemblies. Nuclear Engineering and Design 238(7):1629–1635.

9. **Yiqun Tang**, Zhengdong Cui, Xi Zhang, Shukai Zhao (2008) Dynamic response and pore pressure model of the saturated soft clay around the tunnel under vibration loading of Shanghai subway. Environmental Geology 98(3–4):126–132.
10. Zhengdong Cui, **Yiqun Tang**, Xuexin Yan (2010) Centrifuge modeling of land subsidence caused by the high-rise building group in the soft soil area. Environmental Earth Science 59(8):1819–1826.
11. Zhengdong Cui, **Yiqun Tang**, Xuexin Yan (2010) Evaluation of the geology-environmental capacity of buildings based on the ANFIS model of the floor area ratio. Bulletin of Engineering Geology and the Environment 69(1):111–118.
12. Zhengdong Cui, **Yiqun Tang** (2010) Land subsidence and pore structure of soils caused by the high-rise building group through centrifuge model test. Engineering Geology 113(1–4):44–52.
13. Zhengdong Cui, **Yiqun Tang** (2011) Microstructures of different soil layers caused by the high-rise building group in Shanghai. Environmental Earth Science 63(1):109–119.
14. **Yiqun Tang**, Jie Zhou, Sha Liu, Ping Yang, Jianxiu Wang (2011) Test on cyclic creep behavior of mucky clay in Shanghai under step cyclic loading. Environment Earth Science 63(2):321–327.
15. Ping Yang, **Yiqun Tang**, Nianqing Zhou, Jianxiu Wang, Tianyu She, Xiaohui Zhang (2011) Characteristics of red clay creep in karst caves and loss leakage of soil in the karst rocky desertification area of Puding County, Guizhou, China. Environment Earth Science 63(3):543–549.
16. Xingwei Ren, **Yiqun Tang**, Xu Yiqing, Yuandong Wang, Xi Zhang, Sha Liu (2011) Study on dynamic response of saturated soft clay under the subway vibration loading I: instantaneous dynamic response. Environment Earth Science 64(7):1875–1883.
17. **Yiqun Tang**, Jie Zhou, Jun Hong, Ping Yang, Jianxiu Wang (2012) Quantitative analysis of the microstructure of Shanghai muddy clay before and after freezing. Bulletin of Engineering Geology and the Environment 71(2):309–316.
18. Chunling Yan, **Yiqun Tang**, Yuandong Wang, Xingwei Ren (2012) Accumulated deformation characteristics of silty soil under the subway loading in Shanghai. Natural Hazards 62(2):375–384.
19. Jie Zhou, **Yiqun Tang**, Ping Yang, Xiaohui Zhang, Nianqing Zhou, Jianxiu Wang (2012) Inference of creep mechanism in underground soil loss of karst conduits I. Conceptual model. Natural Hazards 62(3):1191–1215.
20. Ping Yang, **Yiqun Tang**, Jianxiu Wang, Yang Yang, An Xin (2012) Test on consolidation of dredger fill by cube grid of plastic drain board preinstalled. Engineering Geology 127:81–85.
21. Wang Zhiliang, **Yiqun Tang**, Wang Jianguo (2011) Strength and toughness properties of steel fiber reinforced concrete under repetitive impact. Magazine of Concrete Research 63(11):883–891.
22. Jie Zhou, **Yiqun Tang**, Xiaohui Zhang, Tianyu She, Ping Yang, Jianxiu Wang (2012) The influence of water content on soil erosion in the desertification area of Guizhou, China. Carbonates and Evaporites 27(2):185–192.

23. **Yiqun Tang**, Jie Zhou, Xiaojun He, Ping Yang, Jianxiu Wang (2012) Theoretical and experimental study of consolidation settlement characteristics of hydraulic fill soil in Shanghai. Environment Earth Science 67(5):1397–1405.
24. **Yiqun Tang**, Xingwei Ren, Bin Chen, Shoupeng Song, Jianxiu Wang and Ping Yang (2012) Study on land subsidence under different plot ratio through centrifuge model test in soft-soil territory. Environment Earth Science 66(7): 1809–1816.
25. Xingwei Ren, **Yiqun Tang**, Jun Li, Qi Yang (2012) A prediction method using grey model for cumulative plastic deformation under cyclic loads. Natural Hazards 64(1):441–457.
26. Chunling Yan, **Yiqun Tang**, Yuting Liu (2013) Study on fractal dimensions of the silty soil around the tunnel under the subway loading in Shanghai. Environment Earth Science 69(5):1529–1535.
27. **Yiqun Tang**, Jun Li, Xiaohui Zhang, Ping Yang, Jianxiu Wang, Nianqing Zhou (2013) Fractal characteristics and stability of soil aggregates in karst rocky desertification areas. Natural Hazards 65(1):563–579.
28. **Yiqun Tang**, Zhengdong Cui, Xi Zhang (2008) Dynamic response of saturated silty clay around a tunnel under subway vibration loading in Shanghai. The 6th International Symposium geotechnical Symposium Geotechnical Aspects of Underground Construction in Soft Ground. April 2008 in Shanghai:843–847.
29. Zhengdong Cui, **Yiqun Tang**, Xi Zhang (2008) Deformation and pore pressure model of saturated soft clay around a subway tunnel. The 6th International Symposium geotechnical Symposium Geotechnical Aspects of Underground Construction in Soft Ground. April 2008 in Shanghai:769–774.
30. **Yiqun Tang**, Ping Yang, Shen Feng, Nianqing Zhou (2008) Analysis of behavior of melted dark green silty soil. Frontiers of Architecture and Civil Engineering in China 2(3):242–245.
31. **Yiqun Tang**, Jie Zhou, Changqing Luan, Jianxiu Wang, Ping Yang (2010). Analysis on the surrounding land subsidence effect by the trial dewatering in Yishan Road station of Shanghai Metro. 11th Congress of the International Association for Engineering Geology and the Environment Geologically Active, September 2010, Auckland, New Zealand:4517–4524.

Chinese Papers

1. **Tang Yiqun**, Ye Weimin, Zhang Qinhe (1993) Some problems in shield-driven tunneling of Shanghai metro (part 1). Underground Space 13(2):94–99.
2. **Tang Yiqun**, Ye Weimin, Zhang Qinhe (1993) Some problems in shield-driven tunneling of Shanghai metro (part 2). Underground Space 13(2):171–177.
3. **Tang Yiqun**, Ye Weimin, Zhang Qinhe (1994) Study of ground settlement caused by subway tunnel construction in Shanghai. Underground Space 15(4):250–258.

4. Zhang Qinhe, **Tang Yiqun**, Yang Linde (1994) A study on tunneling technology with access to tunnel for shield. Underground Space 14(2):110–119.
5. **Tang Yiqun**, Xu Chao (1997) Problems of environmental geology of shanghai urban development in the 21st century. Underground Space 17(2):95–98.
6. Ye Weimin, **Tang Yiqun**, Yang Linde (1998) Discussion on inspection method of composite foundation of the cement-soil mixed piles. Geotechnical Investigation & Surveying 1:18–20.
7. Ye Weimin, **Tang Yiqun**, Yang Linde, Ye Yaodong (1998) A study of effects of dynamic consolidation for reinforcing saturated soft soil foundation. Rock and Soil Mechanics 19(3):72–76.
8. Ye Weimin, **Tang Yiqun**, Shao Zhongxin (1999) Study of reflected wave form characteristics of rock-socketed pile. Site Investigation Science and Technology 3:63–65.
9. Zhang Shijie, **Tang Yiqun**, Lin Zhi (2000) Probe to the design of the bearing capacity of the rock-socketed cast-in-place pile. Shanghai Geology 2:10–12.
10. Lv Shaowei, **Tang Yiqun**, Ye Weimin (2000) Experimental studies on consolidation law of the shallow soil layer containing high-pressure gas. Chinese Journal of Geotechnical Engineering 22(6):734–737.
11. **Tang Yiqun**, Lin Zhi, Zhao XiaoYun (2000) The finite-element analysis of slope stabilization in the weak soil. Shanghai Geology 4:21–24.
12. **Tang Yiqun**, Ye Weimin, Huang Yu (2001) Full-filled-reviving karst cave – the development characteristics of YiXing Muli cave, Hydrogeology & Engineering Geology 28(5):39–42.
13. Ye Yaodong, **Tang Yiqun**, Ye Weimin, Yin Jianping (2001) Discussion on related problems in construction of SWM working practice. Shanghai Geology 4:62–64.
14. **Tang Yiqun**, Lv Shaowei, Ye Weimin, Huang Yu (2001) Experimental study on consolidation law of unsaturated soil containing high-pressure gas. China Civil Engineering Journal 34(6):100–104.
15. **Tang Yiqun**, Ye Weimin, Huang Yu (2002) Discussion on several problems of construction of deep foundation ditch project, Construction Technology 31(1):5–11.
16. Ye Yaodong, **Tang Yiqun**, Ye Weimin (2002) Discussion of pipe-jacking construction technique in urban area. Construction Technology 31(1):34–35.
17. **Tang Yiqun**, Wang Yanling, Ye Weimin, Zhao Xiaoyun (2002) Unloading protecting for the original water conduit during excavation of foundation pit of Shanghai Science and Technology Building. Hydrogeology & Engineering Geology 29(1):31–34.
18. **Tang Yiqun**, Song Yonghui, Ye Yaodong (2002) Root pile and pressure-grouting applied jointly to project of water supply pile protection. Architecture Technology 33(11):839–840.
19. Huang Yu, **Tang Yiqun**, Ye Weimin, Chen Zhuchang (2002) The research status of pile vertical seismic capacity. Special Structures 19(1):46–49.

20. Zhou Nianqing, **Tang Yiqun**, Lou Menglin (2002) Analysis of regional earthquake characteristics and dynamic parameters of Third Phase Nuclear Power engineering in Qinshan. Journal of Earthquake Engineering and Engineering Vibration 22(6):31–37.
21. **Tang Yiqun**, Shi Weihua, Zhang XiL (2003) The experiments study and theoretical analyses on piping and flow soil. Shanghai Geology 1:25–31.
22. **Tang Yiqun**, Huang Yuu, Ye Weimin, Wang Yanling (2003) Critical dynamic stress ratio and dynamic strain analysis of soils around the tunnel under subway train loading. Chinese Journal of Rock Mechanics and Engineering 22(9):1566–1570.
23. **Tang Yiqun**, Zhou Yushi, Huang Yu, Ye Weimin (2004) Cross-platform calculating program of final settlement of single pile developed with Delphi 6. Hydrogeology & Engineering Geology 31(2):59–63.
24. **Tang Yiqun**, Wang Yanling, Huang Yu, Zhou Zaiyang (2004) Dynamic strength and dynamic stress–strain relation of silt soil under traffic loading. Journal of Tongji University (Natural Science) 32(6):701–704.
25. **Tang Yiqun**, Liu Binyang, Zhao Shukai, Huang Yu (2004) Research on influence of high-pressure marsh gas on sandy silt engineering. Journal of Tongji University (Natural Science) 32(10):1316–1319.
26. **Tang Yiqun**, Song Yonghui, Zhou Nianqing, Huang Yu, Ye Weimin (2005) Experimental research on troubles of EPB shields construction in sandy soil. Chinese Journal of Rock Mechanics and Engineering 24(1):52–56.
27. Ye Weimin, **Tang Yiqun**, Cui Yujun (2005) Measurement of soil suction in laboratory and soil-water characteristics of Shanghai soft soil. Chinese Journal of Geotechnical Engineering 27(3):347–349.
28. Sun Jie, **Tang Yiqun**, Huang Yu (2005) The comparative analysis on internal force between two types of stabilizing pile. Shanghai Geology (2):19–23.
29. **Tang Yiqun**, Zhang Xi, Zhou Nianqing, Huang Yu (2005) Microscopic study of saturated soft clay's behavior under cyclic loading. Journal of Tongji University (Natural Science) 33(5):626–630.
30. **Tang Yiqun**, Zhang Xi, Wang Jianxiu, Song Yonghui (2005) Earth pressure balance shield tunnelling-induced disturbance in silty soil. Journal of Tongji University (Natural Science) 33(8):1031–1035.
31. **Tang Yiqun**, Shen Feng, Hu Xiangdong, Zhou Nianqing, Zou Changzhong, Zhu JH (2005) Study on dynamic constitutive relation and microstructure of melted dark green silty soil in Shanghai. Chinese Journal of Geotechnical Engineering 27(11):1249–1252.
32. **Tang Yiqun**, Cui Zhendong, Yan Chunling, Guo Changqing (2006) Application of little cavity blasting in purple shale ecology construction of central Hunan Province. Bulletin of Soil and Water Conservation 26(2):81–84.
33. **Tang Yiqun**, Shen Feng, Zhang Xi, Hu Xiangdong, Zhou Nianqing, Zou Changzhong (2006) Experimental study of instantaneous uniaxial compressive strength of artificially frozen soils in soft soil area. Journal of Engineering Geology 14(3):376–379.

34. Cui Zhendong, **Tang Yiqun**, Guo Changqing, Yan Chunling (2006) Analysis and calculation for instantaneous flow on single-acting vane pump. Journal of Vibration. Measurement & Diagnosis 26(3):185–187.
35. Yang Ping, **Tang Yiqun**, Peng Zhenbin, Chen An (2006) Study on grouting simulating experiment in sandy gravels. Chinese Journal of Geotechnical Engineering 28(12):2134–2138.
36. Wang Hanmei, **Tang Yiqun**, Yan Xuexin, Wang Jianxiu, Lu Chen (2006) An unequal time-interval GM(1,1) model for predicting ground settlement in Shanghai soft soil engineering. Journal of Engineering Geology 14(3):398–400.
37. Wang Jianxiu, **Tang Yiqun**, Zhu Hehua, Zhou Nianqing, Lu Yongchun, Wang Hongxia (2006) 3D monitoring and analysis of landslide deformation caused by twin-arch tunnel. Chinese Journal of Rock Mechanics and Engineering 25(11):226–232.
38. Zhou Nianqing, **Tang Yiqun**, Wang Jianxiu, Zhang Xi, Hong Jun (2006) Response characteristics of pore pressure in saturated soft clay to the metro vibration loading. Chinese Journal of Geotechnical Engineering 28(12):2134–2138.
39. Wang Jianxiu, **Tang Yiqun**, Wang Hanmei, Lu Chen (2007) Gray analysis on physical simulation of land subsidence caused by high-rising building group. Journal of Tongji University (Natural Science) 35(4):451–454.
40. **Tang Yiqun**, Yang Ping, Shen Feng, Zhou Nianqing (2007) Microscopic study of dark green silty soil behavior after melted. Journal of Tongji University (Natural Science) 35(1):6–9.
41. Zhang Xi, **Tang Yiqun**, Zhou Nianqing, Wang Jianxiu, Zhao Shukai (2007) Dynamic response of saturated soft clay around a subway tunnel under vibration load. China Civil Engineering Journal 40(2):85–88.
42. Cui Zhendong, **Tang Yiqun**, Guo Changqing (2007) Fluid–structure interacted vibration of an element model of parallel-plate fuel assembly. Journal of Vibration and Shock 26(2):48–50.
43. Cui Zhendong, **Tang Yiqun**, Guo Changqing, Yan Chunling (2007) Flow-induced vibration and stability of parallel-plate assembly. Journal of Earthquake Engineering and Engineering Vibration 27(1):86–91.
44. Luan Changqing, **Tang Yiqun**, Yun Jie, Cui Zhendong (2007) Influence of water level of Shuhe hydraulic power station on reservoir bank stability. Journal of Catastrophology 22(1):65–68.
45. **Tang Yiqun**, Yan Xuexin, Wang Jianxiu, Lu Chen (2007) Model test study of influence of high-rise building on ground subsidence. Journal of Tongji University (Natural Science) 35(3):320–325.
46. Wang Yanxiu, **Tang Yiqun**, Zang Yizhong, Chen Jiang, Han Shuangping (2007) Experimental studies and new ideas on the lateral stress in soil. Chinese Journal of Geotechnical Engineering 29(34):430–435.
47. **Tang Yiqun**, Zhang Xi, Zhao Shukai, Wang Jianxiu, Zhou Nianqing (2007) Model of pore water pressure development in saturated soft clay around a subway tunnel under vibration load. China Civil Engineering Journal 40(4):67–70.

48. Cui Zhendong, **Tang Yiqun**, Yan Chunling, Guo Changqing, Zhou Nianqing, Wang Jianxiu (2007) Research on the method of water-clay composite blasting for tunnel. Chinese Journal of Underground Space and Engineering 3(2):248–251.
49. Cui Zhendong, **Tang Yiqun**, Lu Chen, Luan Changqing, Wang Xinghan (2007) Prediction of ground settlement induced due to changes in engineering environment in Shanghai. Journal of Engineering Geology 15(2):233–236.
50. **Tang Yiqun**, Cui Zhendong, Wang Xinghan, Lu Chen (2007) Preliminary research on the land subsidence induced by the engineering environmental effect of dense high-rise building group. Northwestern Seismological Journal 29(2):105–108.
51. Cui Zhendong, **Tang Yiqun**, Guo Changqing (2007) Fluid–structure interaction vibration of the parallel plate fuel assembly with nonlinear support. Northwestern Seismological Journal 22(2):119–122.
52. Luan Changqing, **Tang Yiqun**, Zhao Fasuo, Cui Zhendong (2007) Study on surface subsidence and deformation in mineral region in Hancheng City, Shaanxi Province. Journal of Natural Disasters 16(3):81–85.
53. Yang Ping, **Tang Yiqun**, Wang Jianxiu, Zhou Nianqing, Yan Xuexin (2007) Study on consolidation settlement of dredger fill under deadweight using large strain theory and centrifuge. Chinese Journal of Rock Mechanics and Engineering 26(6):1212–1219.
54. Luan Changqing, **Tang Yiqun**, Gao Wenbin, Xia Yan (2007) Geochemical inverse simulation of Huanhe water-bearing layers in Ordos Cretaceous Basin. Journal of Natural Disasters 16(4)169–173.
55. Cui Zhendong, **Tang Yiqun** (2007) Domestic and international recent situation and research of land subsidence disasters. Northwestern Seismological Journal 29(3):275–278.
56. Luan Changqing, **Tang Yiqun**, Yun Zhengwen (2007) Geological-geochemical characteristics and genesis of Ma'anqiao gold deposit. Contributions to Geology and Mineral Resources Research 22(3):190–194.
57. **Tang Yiqun**, Zhang Xi, Zhao Shukai, Wang Jianxiu (2007) A study on the fractals of saturated soft clay surrounding subway tunnels under dynamic loads. China Civil Engineering Journal 40(11):86–91.
58. **Tang Yiqun**, Luan Changqing, Zhang Xi, Wang Jianxiu, Yang Ping (2008) Numerical simulation of saturated soft clay's deformation around tunnel under subway vibrational loading. Chinese Journal of Underground Space and Engineering 4(1):105–110.
59. Luan Changqing, **Tang Yiqun**, Lin Bin (2008) Study on softening constitutive model of Malan loess. Journal of Chongqing Jianzhu University 30(2):53–56.
60. **Tang Yiqun**, Lv Xilin, Yang Ping (2008) Preliminary analysis of earthquake-induced main geological hazards in Qingchuan Area. Structural Engineers 24(3):20–23.
61. **Tang Yiqun**, Luan Changqing (2008) Analysis of the large indoor model test on water drainage for Yishanlu metro station in Shanghai. Chinese Journal of Underground Space and Engineering 4(3):493–488.

62. **Tang Yiqun**, Luan Changqing, Wang Jianxiu, Zhu Yanfei, Pan Weiqiang (2008) Analysis of the effects of environments for dewatering in a metro station in Shanghai. Journal of Wuhan University of Technology 30(8):147–151.
63. **Tang Yiqun**, Luan Changqing (2008) Settlement analysis of a laboratory model test of dewatering of a metro station in Shanghai. Journal of Chongqing Jianzhu University 30(4):68–72.
64. **Tang Yiqun**, She Tianyu, Zhang Xiaohui, Yang Ping, Wang Jianxiu (2009) Changing of red clay shear strength with water content under rainfall in karst rocky desertification areas, Guizhou Province. Journal of Engineering Geology 17(2):249–252.
65. **Tang Yiqun**, Yang Ping, Wang Jianxiu (2009) Investigation on reformation of curriculum of bachelor's degree and teaching contents. China Electric Power Education 8:104–105.
66. **Tang Yiqun**, Zhang Xiaohui, Zhao Shukai, Wang Jianxiu (2009) Correlatability of microstructure change and macroscopical deformation of soft clay under subway load. Journal of Tongji University (Natural Science) 37(7):872–877.
67. **Tang Yiqun**, Hong Jun, Yang Ping, Wang Jianxiu (2009) Frost-heaving behaviors of mucky clay by artificial horizontal freezing method. Chinese Journal of Geotechnical Engineering, 31(5):772–776.
68. **Tang Yiqun**, Zhao Shukai, Yang Ping, Wang Jianxiu, Zhang Xi (2009) Quantitative analysis of the microscopic behavior of saturated soft clays under cyclic loading. China Civil Engineering Journal, 42(8):98–103.
69. Ren Xingwei, **Tang Yiqun**, Dai Yunxia, Fang Yu (2009) Improved method for calculating landslide initial surge height. Journal of Hydraulic Engineering 40(9):1116–1119.
70. **Tang Yiqun**, Zhang Xiaohui, She Tianyu, Yang Ping, Wang Jianxiu (2009) Wet sieving for stability of brown clayey clay in karst rocky desertification area in Puding county, Guizhou Province. Journal of Engineering Geology 17(6): 817–822.
71. **Tang Yiqun**, Yu Huang (2009) Review of spring protection. Journal of Anhui Agricultural Sciences 37(26):12814–12819.
72. Song Zhigang, **Tang Yiqun**, Hong Jun, Wang Yuandong, Li Renjie, Zhao Hua (2009) Experiment study on unconfined compressive strength before and after thaw of the forth layer silt clay in Shanghai. Low Temperature Architecture Technology (12):85–86.
73. Song Zhigang, **Tang Yiqun**, Deng Bo, Jin Linzhi (2010) Study on rule of co-acting between externally prestressed concrete beam and external tendons. Journal of the China Railway Society 32(1):65–72.
74. Yan Chunling, **Tang Yiqun** (2010) Improved construction of smooth blasting and support of anchoring and shotcreting on tunnel engineering. Mining and Metallurgy 19(1):4–6.
75. Yan Chunling, **Tang Yiqun** (2010) Research and application of reasonable blasthole depth to tunnel evaluation operation. Engineering Blasting 16(1):6–8.

76. **Tang Yiqun**, Zhang Xiaohui, Zhou Jie, She Tianyu, Yang Ping, Wang Jianxiu (2010) The mechanism of underground leakage of soil in karst rocky desertification areas–a case in Chenqi small watershed, Puding, Guizhou Province. China Karst 29(2):121–126.
77. **Tang Yiqun**, Wang Yuandong, Li Renjie, Zhao Hua (2010) A study on fractals of reinforced soft clay around tunnel under subway vibrational loading. Journal of Tongji University (Natural Science) 38(7):1002–1006.
78. Wang Yuandong, **Tang Yiqun**, Liao Shaoming, Li Renjie (2011) Experimental of pore pressure of reinforced soft clay around tunnel under subway vibrational loading. Journal of Jilin University (Earth Science Edition) 41(1):188–194.
79. Yan Chunling, **Tang Yiqun**, Wang Yuandong, Li Renjie (2011) Fuzzy orthogonal analysis on pore water pressure of reinforced soft clay under cyclic loads. Journal of Jilin University (Earth Science Edition) 41(3):805–811.
80. Yan Chunling, **Tang Yiqun** (2011) Advances in researches on dynamic properties of silty soil under subway loading. Northwestern Seismological Journal 33(2):200–205.
81. Yan Chunling, **Tang Yiqun**, Liu Sha (2011) Accumulative deformation characteristics of saturated soft clay under subway loading in Shanghai. Journal of Tongji University (Natural Science) 39(7):977–978.
82. **Tang Yiqun**, Zhao Hua, Wang Yuandong, Li Renjie (2011) Characteristics of strain accumulation of reinforced soft clay around tunnel under subway vibration loading. Journal of Tongji University (Natural Science) 39(7):972–977.
83. Song Shoupeng, **Tang Yiqun** (2012) Best floor area ratio of high density buildings based on rheological theory. Journal of Central South University (Science and Technology) 43(6):2349–2356.
84. Xu Yiqing, **Tang Yiqun**, Ren Xingwei, Wang Yuandong (2012) Experimental study on dynamic elastic modulus of reinforced soft clay around subway tunnel under vibration loading. Engineering Mechanics 29(7):250–269.

Index

A
Attenuation, 5, 9, 32–34, 38, 41, 43, 58–59, 66, 94, 95, 185
Average relative error, 209, 210, 213
Axle loads, 6, 9, 30, 37, 59, 221

C
Clay
 microstructure, 13–16, 18, 19, 115, 117, 130, 147–152, 225, 227, 228
 minerals, 14, 16, 49, 102, 108, 114–116, 121, 151, 226
 sensitivity, 105
Coefficient of Skempton, 3
Compressibility, 2, 27, 42, 46, 60, 62, 63, 74, 108, 109, 111, 115, 119
Confining pressure, 11, 12, 44, 45, 48, 61, 62, 64, 66, 68–71, 74, 75, 88, 89, 98, 226
CSR. *See* Cyclic stress ratio (CSR)
Cumulative plastic strain, 205, 206, 210, 211, 216
Cyclic stress ratio (CSR), 10, 12, 15, 18, 42, 62, 65–68, 70, 71, 81–87, 98, 125, 138, 140, 145–150, 152, 204, 226, 227
Cyclic triaxial tests, 2, 10–13, 15, 17–19, 41, 43, 44, 49, 60, 62, 63, 67, 70, 73–76, 87, 91, 97, 102, 126
Cylindrical pore model, 128

D
Damage mechanism, 18, 19, 41, 42, 60, 98
Dark green stiff clay (DGSC), 27, 181

E
Earth pressure, 5, 9, 19, 26, 31, 38, 122, 123, 140, 147, 195, 225
Eccentric wheel, 32, 170–171
Electric double layer, 49–52
Excess pore pressure, 26, 35, 42, 54, 55, 57, 59, 94, 95, 98, 192, 219, 226
ExpDecay2 model, 95

F
Field tests, 5–8, 10, 19, 25–38, 45, 56, 75, 122, 147, 156, 169, 173, 181, 184, 186–189, 220, 225, 228
Finite element modeling, 155–189
Fourier series, 4, 9
Fractal dimension, 14–16, 103, 104, 140–148, 152, 227, 228
Fractal model, 140–147

Deformation mechanism, 15, 16, 19, 20, 53, 60, 94, 98, 102, 106, 114, 124, 126, 152, 204, 225, 227, 228
Dynamic constitutive relation, 158–159
Dynamic cyclic triaxial test, 18, 19, 41, 43, 74
Dynamic elastic modulus, 41, 68, 89, 97
Dynamic finite element analysis, 161–166
Dynamic loading, 4, 12, 17, 42, 64, 65, 70, 88, 101, 166, 167, 169–176, 191, 225
Dynamic pore water pressure, 8, 19, 42, 64, 158, 161, 166, 219
Dynamic properties, 2, 7, 9–13, 19, 36, 41, 60, 62, 68, 70, 98

G

Global digital systems (GDS), 2, 18, 19, 41, 43–45, 48, 61, 74, 90–92, 95, 130, 136
Gray prediction theory, 203–219, 222, 227
Grouting pre-settlement, 198–199
Grouting quantity, 195–198

H

High frequency, 30, 31, 45, 66, 98, 173, 225

L

Liquid limit, 2, 52, 108, 109
Logistic model, 94, 98
Long-term additional settlement, 17, 202–203
Long-term settlement, 16–19, 191, 192, 200–222, 227
Low frequency, 30, 31, 45, 66, 68, 173, 185, 225

M

Mean distribution radius, 132, 139, 147, 148, 150, 152, 227
Mercury intrusion porosimetry (MIP), 2, 18, 19, 41, 97, 103, 104, 127–132, 138, 142, 145, 151, 225
Mesh generation, 179–184
Microstructure, 2, 13–16, 18, 19, 42, 49, 60, 98, 101–152, 225, 227, 228
Microstructure basic element, 115–117
MIP. *See* Mercury intrusion porosimetry (MIP)
Model checking, 209–210
Mucky clay, 2, 11, 18, 27, 29, 30, 43, 46, 63, 71, 74, 76, 78, 79, 81, 83, 84, 86–89, 106, 108–114, 119, 121, 126, 136, 138, 151, 152, 177, 181, 184, 192, 195, 212, 226, 227

N

National Natural Science Foundation, 2
NNGM (1, 1), 192, 209–219, 221, 227
Non-isochronous data sequence, 207, 219, 221
Numerical method, 156, 204

P

Permeability, 2, 3, 27, 42, 44, 46, 60, 74, 107–109, 111, 115, 119, 122, 131, 160, 202

Plastic limit, 52
Pore surface area, 128, 141
Pore water development, 53–58
Porosity, 14, 15, 18, 19, 53, 103, 104, 108, 109, 111, 119, 120, 122, 124, 128, 130, 131, 139, 140, 146, 147, 149, 150, 152, 227

R

Rail irregularity, 4
Random vibration theory, 4
Residual mercury volume, 132
Retention factor, 124, 131, 132, 139, 148–152, 227

S

Scanning electron microscopy (SEM), 2, 13–15, 18, 41, 102–104, 106, 107, 109, 114, 117, 119–124, 126, 127, 138, 151, 225
Settlement prediction, 15, 17, 18, 20, 191–222, 226
Settlement tank, 194–195
Shear strength, 2, 12, 17, 42, 51, 71–73, 108, 109, 112, 115
Small error probability, 209, 210, 213
Soil expansibility, 105–106
Soil structure, 3, 13, 14, 26, 36, 52, 68, 69, 95, 97, 101–103, 105, 106, 109, 112–114, 122, 125, 126, 152, 201, 202, 225, 227
Standard deviation ratio, 210, 213
Static-dynamic analysis, 166–169
Stereo space traffic system, 1
Stress-strain properties, 53

T

Threshold critical stress ratio, 202
Track irregularity, 4, 6, 32, 169–173, 221
Transient analysis, 157, 186–189
Tunnel, 2, 25, 41, 122, 156, 191, 225
Tunnel axis, 32–34, 38, 75, 124, 126, 168, 178, 192–194, 196–198, 200, 201, 221, 225

U

Urban transportation, 1, 25, 73

V

Vibration
    cyclic number, 58, 65, 93, 98, 112
    frequency, 5, 10, 19, 62, 64, 68, 226

Index

loading, 2–6, 8, 16–18, 41, 60, 63, 65, 69, 71, 75, 87, 91, 92, 94, 98, 104, 131
response wave, 31
Void ratio, 2, 42, 46, 60, 63, 74, 87, 108, 109, 121, 138, 151, 213, 226

W

Water content, 2, 14, 42, 46, 52, 60, 107–109, 111

Wheel flat, 170, 221
Wheel-rail contact, 4, 6, 32

X

X-ray diffraction (XRD), 13, 101, 102, 114–116

Y

Yield stress, 3, 12, 13, 168

Printed by Publishers' Graphics LLC
DBT140111.15.16.9